T0340175

PROCESS SAFETY MANAGEMENT AND HUMAN FACTORS

PROCESS SAFETY MANAGEMENT AND HUMAN FACTORS

A PRACTITIONER'S EXPERIENTIAL APPROACH

Edited by

WADDAH S. GHANEM AL HASHMI FEI, AFIChemE, FIEMA

Senior Director, Emirates National Oil Company (ENOC), Dubai, United Arab Emirates
Hon. Chairman, Energy Institute-Middle East (EI-ME), Dubai, United Arab Emirates

Butterworth-Heinemann
An imprint of Elsevier

Butterworth-Heinemann is an imprint of Elsevier
The Boulevard, Langford Lane, Kidlington, Oxford OX5 1GB, United Kingdom
50 Hampshire Street, 5th Floor, Cambridge, MA 02139, United States

Notices
Knowledge and best practice in this field are constantly changing. As new research and experience
broaden our understanding, changes in research methods, professional practices, or medical
treatment may become necessary.

Practitioners and researchers must always rely on their own experience and knowledge in
evaluating and using any information, methods, compounds, or experiments described herein.
In using such information or methods they should be mindful of their own safety and the safety of
others, including parties for whom they have a professional responsibility.

To the fullest extent of the law, neither the Publisher nor the authors, contributors, or editors,
assume any liability for any injury and/or damage to persons or property as a matter of products
liability, negligence or otherwise, or from any use or operation of any methods, products,
instructions, or ideas contained in the material herein.

British Library Cataloguing-in-Publication Data
A catalogue record for this book is available from the British Library

Library of Congress Cataloging-in-Publication Data
A catalog record for this book is available from the Library of Congress

ISBN: 978-0-12-818109-6

For Information on all Butterworth-Heinemann publications
visit our website at https://www.elsevier.com/books-and-journals

Publisher: Susan Dennis
Acquisitions Editor: Anita Koch
Editorial Project Manager: Redding Morse
Production Project Manager: Chandramohan, Paul Prasad
Cover Designer: Mark Rogers

Typeset by MPS Limited, Chennai, India

Working together
to grow libraries in
developing countries

www.elsevier.com • www.bookaid.org

Dedication

This book is dedicated to everyone who have been adversely affected by process safety incidents throughout time. It is dedicated also to all those practitioners who have been and still trying very hard to prevent process safety incidents and understand better how to manage effective strategies for their prevention; and through collaboration or otherwise taking the time to better understand the human element and factors that underlie so many process safety failures that have caused incidents leading to loss of life, assets and damage to the environment.

For and on behalf of all the authors

Dr. Waddah S. Ghanem Al Hashmi, AFIChemE, FEI, FIEMA

Contents

12. Prestart-up and shutdown safety reviews   147
CHITRAM LUTCHMAN AND RAMAKRISHNA AKULA

13. Contractor management   159
CHITRAM LUTCHMAN AND RAMAKRISHNA AKULA

14. Emergency response management and control   179
AHMED KHALIL EBRAHIM

15. Human performance within process safety management compliance assurance   201
JANETTE EDMONDS

16. Regulating PSM and the impact of effectiveness 213
WADDAH S. GHANEM AL HASHMI AND HIRAK DUTTA

17. Readying the organization for change: communication and alignment 221
CHITRAM LUTCHMAN AND RAMAKRISHNA AKULA

18. Do we really learn from loss incidents? 235
RAM K. GOYAL

About the authors

Dr. Waddah S. Ghanem Al Hashmi
BEng (Hons), MBA, MSc, DipSM, DipEM, AFIChemE, FEI, MIoD

Chief Editor
Graduated from the University of Wales College Cardiff, School of Engineering, with a Bachelor of Engineering (Honours) in Environmental Engineering. Waddah is considered one of the global authorities on Governance and leadership in EHS.

Currently, he is Senior Director for Sustainability, Operational, and Business Excellence for the ENOC Group and overseas all the sustainability, CSR, energy and resource management, and reporting as well as business and operational excellence, innovation and process improvement, and LSS projects. He has spent the last 22 years from being a consultant to an EHS supervisor at the refinery to assistant EHS advisor in the group, later to grow the EHS compliance function until he became director of EHSQ compliance in January 2010.

He was appointed in 2015 as executive director of EHSSQ and corporate affairs and in June 2018 was given the role of senior director for sustainability, operational and business excellence. He also chairs various committees in ENOC including the governance and oversight for the SAP digital transformation, the operational excellence framework committee, and the asset integrity committee.

Waddah holds several academic and vocational qualifications including PG diplomas an and an MSc in Environmental Sciences; an executive MBA and doctorate from the University of Bradford in the United Kingdom in 2014/2015. He was appointed as adjacent lecturer in the Universiti Techmologi Petronas (UTP) in 2018, and he is part of the industrial advisory committee for the Herriot Watt University in the United Arab Emirates and is an external examiner for the University of Petroleum Environmental Sciences (UPES) in Dhurdun, India.

His publications include various research as well as practitioner journals and conference papers. This is in addition to seven internationally published books, mainly in safety management, operational excellence, reflective learning, hse practitioner executive development, and governance and leadership as well as theology. Two books are currently under development, one in project management safety in mega projects, and another on governance and leadership in health and safety.

He served as Vice Chairman of Dubai Carbon (DCCE PJSC) from 2010 to 2019; he is a Board Member of the not-for-profit Emirates Environment Group; an Executive Committee Member (Board) of the Oil Companies International Marine Forum (OCIMF) and a Board/ Council member of the Energy Institute (EI) in the United Kingdom and serves as the Hon. Chairman of the Energy Institute Middle East Branch. He is also the senior advisor to the Board of the Clean Energy Business Council. He was born and is based in Dubai, United Arab Emirates.

Hari Kumar Polavarapu
BSc, MSc

Hari has over 35 years of experience in the oil and gas industry and managed process, operations, asset integrity, quality, and HSE functions across the oil and gas supply chain. He currently the ENOC Group's EHS Director. His previous work experience includes around eight years as the Director, HSSEQ and Sustainability of Cairn India Ltd based in Gurgaon, India. He had also served in the past reputed companies such as EPPCO (A Chevron JV) and Indian Oil Corporation Ltd in various capacities and functions.

Hari holds a master's degree in chemical engineering with environmental majors from IIT-Chennai, and a bachelor's degree in chemical engineering from Osmania University, Hyderabad. Hari is also a city and guilds certified assessor for NVQ level-3, certified ISO 9001, and ISO 14001 lead auditor and RSoH-certified HACCP auditor.

Hari is quite active in international HSE professional forums and was the founder Chairman of the Global HSE Conference which has now become one of the prestigious biennial HSE events in the world. He has published articles in various industry/HSE journals and presented papers at numerous international forums. He also served as an industry member on the steering committee of the Oil Industry Safety Directorate (India) from 2010 to 2015.

Hari is married and blessed with two children and currently lives in Dubai, United Arab Emirates.

Ahmed Khalil Ebrahim
MSc

Director for EHS at the Bahrain Petroleum Company (BAPCO) and the Director for EHS on the BAPCO Refinery Expansion Project.

Ahmed Khalil has over 36 years of process safety and fire prevention experience in the oil refining industry. He was educated in the United Kingdom and received a MSc in Risk Management and Safety Technology from Aston University. He is currently the Director for EHS at the Bahrain Petroleum Company (BAPCO) and the Director for EHS on the BAPCO Refinery Expansion Project.

Ahmed is a member of the Bahrain Health and Safety Council, Bahrain Health and Safety Standardization Committee, the Society of Petroleum Engineers (SPE) EHS Committee, and the BAPCO Environment and Health and Safety Committee. Internationally he is a board member with the RW Campbell Institute in the US, the GCC National Oil Companies EHS Committee, and the SPE HSE Committee. Ahmed chairs the committee responsible for reviewing and updating Bahrain H&S standards.

Ahmed has many accreditations and vocational qualifications in PSM, HAZOP, Taproot Investigations, OHSAS 18001 as well as OSHA based qualifications. Ahmed published several papers and many presentations covering HSE issues and presented many at national and international conferences and safety events.

Ahmed has some exceptional experience in developing Operational Excellence Systems and integrating PSM into the company management systems. He has an impressively long list of achievements at BAPCO, in Bahrain, and internationally.

He is married with children and lives in the Kingdom of Bahrain.

Dr. Ian Randle
BSc, MSc, PhD, MCIEHF, C.ErgHF

Managing Director of Hu-Tech Human Factors Consultancy, Past-President of the Chartered Institute of Ergonomics & Human Factors (2016—17).

Ian is Managing Director of Hu-Tech Human Factors Consultancy and President of the Chartered Institute of Ergonomics and Human Factors (2016—17). He is a chartered ergonomist and human factors specialist and has over

32 years' experience in human factors consultancy, research, and teaching.

Ian has worked in a broad range of sectors including manufacturing, defense, aviation, nuclear, pharmaceuticals, and health care. However, over the past 17 years Ian has mostly worked in oil and gas major projects. His work has ranged from managing the human factors integration process through the stages of a project, to undertaking human error analysis of safety critical tasks. As the MD of Hu-Tech he has overseen the Human Factors support in over 100 oil/gas and maritime capital projects around the world.

Dr. Chitram (Chit) Lutchman
BSc, MBA, DBA, 1st Class Power Eng, CSP, CRSP

Dr. Chit Lutchman is a hands-on Oil and Gas professional with more than 30 years of experience in leading operations management, project management, EH&S, PSM, and OEMS. Chit has been the lead author in published books in the areas of project management, EH&S, PSM and OEMS. He was also the lead author of *7 Fundamentals of Operationally Excellent Management System*. In his work, Dr. Lutchman seeks to bring business intelligence, operations discipline, risk management, reliability and EH&S under a single umbrella for sustainable excellence in business performance.

Dr. Lutchman is an internationally recognized EH&S and Management System professional and is has presented at international conferences in the United States, Middle East, Asia and the Caribbean. A consummate learner, Dr. Lutchman challenges himself and others to continually improve. Dr. Lutchman holds a DBA, MBA and a BSc Agricultural Sciences. He is also a first Class Power Engineering License holder and is a certified safety professional (CSP) and a Canadian registered safety professional (CRSP).

Dr. Lutchman has provided solutions for improving operations discipline and managing operations risk and EH&S solutions to ENOC, BAPCO, and North American Oil and Gas majors. Today, Dr. Lutchman is regarded among the world leading experts in EH&S and OEMS.

Today, Dr. Lutchman is regarded as a global expert in HSE and has presented on various HSE and OEMS conferences in Canada, United States, the Middle East, Asia, and the Caribbean.

Janette Edmonds
BSc(Hons), MSc, CErgHF, FCIEHF, CMIOSH

Director and Principal Consultant Ergonomist and Human Factors Specialist, Keil Centre, United Kingdom

Janette has an academic background in psychology and ergonomics. She began her practitioner career in 1994, and since then has gained experience within a wide range of industries including chemical processing, oil and gas, transport systems, manufacturing and production, consumer product design, emergency services, defense, and other applications.

Janette specializes in human factors in engineering and design, human factors in incident investigation, human error analysis, human factors in procedure development, and safety culture assessment. Janette is the course director for the IChemE's Human Factors in Health & Safety program (in United Kingdom and Europe).

She is a fellow of the Institute of Ergonomics and Human Factors (FIEHF) and a chartered member of the Institution of Occupational Safety and Health (CMIOSH). In 1999 she was presented with the William F. Floyd Award for an outstanding contribution to ergonomics.

Dr. Ramakrishna Akula
BEng, PG(Safety), MBA, PhD
Vice Presudent—Operations, Safety Erudite Inc.

Dr. Rama is passionate about business/operations excellence and is committed to helping clients achieve their elusive goals of operations discipline and operations excellence. Dr. Rama is a mechanical engineer with 20 years of experience working for Oil and Gas organizations in North America, Europe, the Middle East and Asia.

Dr. Rama gained significant recognition as a subject matter expert (SME) in plant management, QHSE, organizational strategic performance, operational excellence (HSE), and business excellence (quality) managements systems.

Dr. Rama completed his PhD in material safety with a focus on improving quality and safety around LP Gas cylinders. He is a chartered member in IOSH, UK (CMIOSH), holding the title occupational safety and health practitioner and a chartered mechanical engineer from

Institution of Engineers (India). He also holds an MBA in operations, a postgraduate diploma in HSE and a certificate in disaster management.

An experienced management systems assessor, he has been a lead auditor for recipients of prestigious industry excellence awards in the United Arab Emirates for the implementation of corporate excellence programs. Also, Dr. Rama is the recipient of the United Arab Emirates National Quality Award from ASQ Middle East and North Africa and holds a world record on research publications.

Ram Goyal
Consultant, Bapco Modernization Project (BMP),
Bahrain Petroleum Company

Prior to being appointed to the BMP project, Ram was the Risk Management Advisor and Leader of Central Reliability Engineering at Bahrain Petroleum Company (BAPCO). His areas of interest are hazard and risk analysis, process safety management, environmental risks, contingency planning, insurance, and engineering standards. Prior to joining BAPCO, he was at Fluor Daniel, Inc., Sugar Land, Texas, where his work included reliability engineering, quantitative risk assessment, and HAZOP studies for client companies in the United States, Canada, Mexico, Venezuela, South Africa, and Thailand.

Ram had performed pioneering work in quantitative risk management and loss prevention engineering during a ten-year period with Saudi Aramco in the 1980s. Some of this work was published in the ASSE "Professional Safety" and in Hydrocarbon Processing. His recent publications include papers published in Hydrocarbon World, UK IChemE's the Chemical Engineer and the Loss Prevention Bulletin, and the conference proceedings of the Society of Petroleum Engineers.

He holds an MSc degree in management science from Imperial College, London, and a BTech degree in mechanical engineering from the Indian Institute of Technology, Delhi, India. He is a Fellow of the UK Institution of Mechanical Engineers (FIMechE), a Chartered Engineer (CEng) in the United Kingdom, a registered European Engineer (Eur Ing) in the EEC countries, a Certified Safety Professional (CSP) in the United States, and a professional member of American Society of Safety Professionals.

Dr. Mark Hodgkinson
PhD (App. Chem), BAppSc Chem
Manager, Operational Excellence, Bapco, Bahrain
Petroleum Company

Dr. Mark Cameron Hodgkinson is an operational excellence (OE) expert and scientist with more than 25 years postgraduate experience in operational oil refineries within Australia, Thailand, Japan, and Bahrain. He is a specialist in the identification, analysis, development, and implementation of practical and sustainable continuous improvement initiatives for a wide range of business systems and applications. However, core OE critical systems like PSM are his passion.

He holds a PhD in applied chemistry and his main efforts have been in developing operational intelligence processes to enable assurance. He is published in several disciplines from applied microbiology and chemistry to chemometrics. His areas of expertise include: (1) process and personal safety, especially barrier thinking, BBS, and PSM assurance; (2) quality and process improvement including operational excellence, governance and intelligence, project management, change management, QMS auditing, and business planning, as well as the improvement of design and implementation systems including ISO 9001:2000, Six Sigma, and various management systems and work processes; (3) leading the development of continuous improvement projects to enhance performance of continuous processing plant, with an inherent ability to translate technical information for nontechnical users across multiple business functions; (4) development, management, and mentoring of technical teams, providing role clarity that empowers individuals to meet their goals and deliver successful projects through to handover, with optimal commercial and safety performance; (5) develops robust relationships with cross-functional interfaces, enabling clear understanding of commercial business objectives, and delivering successful strategies that provide high return on investment; (6) leveraging operational intelligence technologies to sustain operational excellence and quality management systems; and (7) chemometrics (chemical analysis statistics, signal analysis, and multivariate processes monitoring), environmental analytical chemistry, applied microbiology, behavioral bioassay, data/information visualization, and scientific research.

Hirak Dutta

Head of Safety and Adviser, Nayara Energy
Limited and Former Executive Director, OISD,
Ministry of Petroleum and NG, India

Hirak Dutta has over 40 years of rich and varied
experience in oil and gas industry in process
design and engineering, operations, trouble
shooting, project management, safety management, HRM at various refineries, refinery HQ
and corporate office of Indian Oil Corporation
Limited.

In the initial stage of career, Hirak was associated with the commissioning of Mathura Refinery. Later, Hirak was the Technical Head of
Haldia Refinery, IOCL. From 2011 until late 2015, Hirak was leading Oil
Industry Safety Directorate (OISD) as executive director, technical directorate under Ministry of Petroleum & Natural Gas, Government of
India, and was responsible for overseeing the safety implementation in
the entire hydrocarbon sector for public and private enterprises. The
responsibility included development of safety standards, risk assessments, accident investigations, safety performance measurement, etc.

Hirak undertook training, seminars, and conferences within the coun-
try and abroad including a condensed MBA program (MEP) at the
Indian Institute of Management, Ahmedabad; Senior Management
Program at Management Development Institute; and World Petroleum
Congress at Calgary, etc.

Hirak authored many papers in technical and safety management
areas which have been published prestigious journals. Hirak is also a
regular faculty in technical, process safety, and project management,
and has presented papers in national/international conferences besides
taking several sessions at academic Institutes like Indian Institute of
Technology (IIT), Indian Institute of Management (IIM), AIChE, etc.

Since November 2015, Hirak is working at Nayara Energy Limited as
head of safety and adviser.

Hirak is a chemical engineer by profession from Jadavpur University.

Associate Prof Dr. Risza Rusli
Associate Professor in Chemical Engineering Department, University Teknologi PETRONAS (UTP)

Dr. Risza Rusli is Associate Professor in Chemical Engineering, University Teknologi PETRONAS (UTP), and Associate Member in Process Safety of IChemE UK. Currently, she is the head of the Centre of Advanced Process Safety, a research and service center at UTP.

Her current research and projects are focusing on development of inherent safety measuring tools for process design, consequence, and risk analysis using computational fluid dynamic simulation. She is also currently involved in development of training and assessment modules to enhance human performance in process industries. In addition, she is actively providing professional courses in process safety topics for local companies in Malaysia.

Dr. Mardhati Zainal Abidin
Postdoctoral Researcher Centre of Advanced Process Safety (CAPS), Universiti Teknologi PETRONAS

Mardhati Zainal Abidin is Postdoctoral Researcher at the Centre of Advanced Process Safety (CAPS), Universiti Teknologi PETRONAS.

She is a member of Board of Engineers Malaysia. Her research interest focuses on development of measurement tool for inherently safety design and development of health risk assessment tool for nanomaterials.

Professor Dr. Azmi Mohd Shariff
Director of Institute of Contaminant Management for Oil and Gas, Universiti Teknologi PETRONAS (UTP)

Azmi Mohd Shariff is Professor in Chemical Engineering Department and Director of Institute of Contaminant Management for Oil and Gas at Universiti Teknologi PETRONAS (UTP), Perak, Malaysia. He was conferred a PhD in Chemical Engineering from University of Leeds, United Kingdom (1996).

He was with Universiti Kebangsaan Malaysia in 1989 as a tutor and was appointed as a lecturer in 1996 upon his return from University of Leeds. He joined UTP in 1997. He was the Head of Industrial Internship in 1998, Head of Chemical Engineering Department from 1999 until 2003, and Director for CO2 Management MOR from 2011 until 2013. Currently, he is the member of Research Centre for CO2 Capture (RCCO2C) and Centre of Advanced Process Safety (CAPS).

He is the author and coauthor of over 170 journal publications and presented more than 120 papers in conferences. He has 18 patents filed, 2 granted patents, 4 copyrights, 3 trademarks, 2 commercialization, and won 11 awards in exhibitions. He led and completed 18 university, 18 government and 14 industrial research grants. He had completed 11 quantitative risk assessment (QRA) projects for industries and conducted 6 short-courses on QRA for industrial professionals. His current research works are in process safety and CO2-NG separation.

In process safety, his research works are in QRA, process safety management (PSM), human factor (HF), and inherent safety (IS).

Dr. Muhammad Athar
Teaching Graduate Assistant at Universiti Teknologi PETRONAS (UTP), Perak, Malaysia

Muhammad Athar has recently finished his PhD in Chemical Engineering and serving as teaching graduate assistant at Universiti Teknologi PETRONAS (UTP), Perak, Malaysia. He has hands-on experience in research and has published several peer-reviewed journal and conference articles and a book chapter.

After graduation in 2007, he has served in Pakistan as a professional chemical engineer for about seven years in versatile capacities. He has worked out in multiple dimensions of the petrochemical industry, that is, shift engineer of an ammonia plant, process engineer, project engineer, and process safety engineer.

He has obtained numerous professional qualifications at the national and international level such as certified energy manager, process safety management, NEBOSH IGC, etc. He has also served as course tutor for process simulation software training across fertilizer sector of Pakistan. His research interests include process safety, risk assessment, gasification, coal and biomass, and modeling and simulation. His professional areas of expertise are regarding process safety, process engineering, and project engineering such as MOC, PHA, PSSR, ERP, consequence analysis and incident investigation/reporting, Dupont process and personnel safety management guidelines, process simulation, energy conservation, and feasibility preparation.

Professor Ron McLeod

BSc (Hons), MSc, PhD, FCIEHFIndependent Human Factors Consultant

With an honors degree in Psychology, an MSc in Ergonomics, and a PhD in Engineering and Applied Science, Ron McLeod has more than 35 years' experience in safety critical and high hazard industries. He is Fellow of the Chartered Institute of Ergonomics and Human Factors (CIEHF) and holds positions as Honorary Professor of Engineering Psychology at Heriot-Watt University, Edinburgh, and Visiting Professor of Human Factors at Loughborough University, School of Design.

Ron has taken an active, and often a leadership, role in human factors initiatives by organizations including the International Association of Oil and Gas Producers (IOGP), the Society of Petroleum Engineers (SPE), the UK's Process Safety Advisory Committee (PSAC), the Energy Institute (EI), the Centre for Chemical Process Safety (CCPS), and the Chartered Institute of Ergonomics and Human Factors (CIEHF).

Ron has been Trustee and Member of Council of the CIEHF and has served on the Board of Directors of the SPE's Human Factors Technical Section. He has been an SPE Distinguished Lecturer and has served as a committee member for the US National Academy of Sciences, Engineering, and Medicine. His first book *Designing for Human Reliability: Human Factors Engineering in the Oil, Gas and Process Industries* was published in 2015.

Preface

For many years, practitioners have debated the effectiveness of process safety management (PSM) systems as opposed to integrated health and safety or otherwise even the occupational health, safety, and environment systems. Clearly, each type of system was developed to address a particular requirement or try to manage a particular set of risks, or otherwise to enhance the risk management processes within an organization.

It is essentially for those reasons that PSM was developed initially as a standard to prevent the process safety catastrophes like the Bhopal, India, incident, which claimed so many lives and destroyed many others. In fact, the devastating impacts on people and their families remain to this day. But is it the management system that can address these issues completely and comprehensively? If so, why are we still having major incidents in the process, oil and gas, petrochemical, and manufacturing industries?

We believe that with every development of a process, plant, equipment, or operation will come some certain risks. Some of these risks are significant and others less so, but regardless of the magnitude and the types of risks that can arise, almost all risks are manageable, and thus almost all, other than natural calamities, are controllable and preventable.

Since the initial inception and implementation of the PSM system as prescribed in the specifically US Occupational Safety and Health Administration (OSHA) PSM standard 29 CFR 1910.119 to the many different standards that followed in the 3 decades that followed many organizations including for example the CCPS and in more recent years the Energy Institute have been trying to enhance PSM standards. We know also that various organizations have adopted and furthermore adapted the PSM systems to integrate or augment their Environment, Health and Safety (EHS) Management Systems as well. We believe that the reason for the development of more PSM standards with newer and reengineered elements is that the standards have yet to reach a more comprehensive level which addresses all the risks effectively.

With experience the industry experts have continued to challenge the standards and elements and tried, through experiential knowledge, especially after incidents, to understand what seems to be still going wrong. Clearly, the safety culture maturity within organizations is a factor to consider. We have also, now with greater research from informed practitioners and human factors specialists, a greater understanding of the importance of strategies of implementing PSM, taking into account the human element.

It is for that reason that this book brings together more than 14 international EHS, PSM, and human factor experts and specialists to address each of the major elements of PSM. Yet the approach is more practitioner and experientially based where the discussion steers away somewhat from the more foundational discussion on the pure theoretical aspects of PSM and attempts to address more practical aspects and the human factors that impact its effective implementation.

What it also tries to do is engage in a discussion on the practical issues with the implementation and while it focuses on the foundational elements of PSM, it draws on EHS management systems, operational excellence, enterprise risk management, managing PSM in major projects, and so on. It focuses also on the serious challenges in implementation and illustrates with various case studies as well.

We believe this book presents a unique value proposition to the practitioner as it provides insightful chapters from a unique and diverse set of authors working as practitioners in organizations, advisors, academics engaged with research, and teaching both PSM and human factors as well as consultants, etc. It provides an experiential approach toward looking at PSM and human factors, which delivers the book's unique value proposition.

A Note on the Structure of this Book

Chapter 1 offers a short history of the development of *Process Safety Management* and provides a short introduction of PSM in a practical context. Chapter 2 is a foundational chapter that talks about *Introduction to Human Factors and the Human Element* and connects the same with PSM. The chapter introduces the importance of the relationship between both areas. In Chapter 3, we address *Leadership in PSM* and the importance of leadership as a fundamental element in driving the application of PSM systems.

Chapter 4 addresses the *Awareness of Risk and the Normalization of Deviation* in a very thought provoking discussion with some simple but powerful case studies. This is followed by Chapter 5, which focuses on *Competency Assurance and Organizational Learning* in the context of human factors and PSM. Chapter 6 addresses the *Integration of Human Factors in Hazard identification and Risk Assessment*. This is then followed by Chapter 7, which addresses the *Inherent Safety Impact in Complying Process Safety Regulations and Reducing Human Error* with some very foundational work on this important aspect of PSM. Chapter 8, *Asset and Mechanical Integrity Management in PSM*, is followed by Chapter 9, which addresses one of the most important PSM elements, *Management of Change (MOC)*. Chapter 10 addresses *Management of Risk through Safe Work Practices* while Chapter 11 addresses a subject many practitioners struggle to appreciate in PSM systems, which is *Process Safety Information (PSI), Hazard Control, and Communication*.

Prestart-Up and Shutdown Safety Reviews is the title of Chapter 12, which is another fundamental PSM element, whereas Chapters 13 and 14 address *Contractor Management* and *Emergency Response Planning*, respectively. Chapter 15 looks at *Human Performance within PSM Compliance Assurance*, which is a unique chapter that provides an insight into why humans fail, the gaps within PSM assurance, and what can be done to assure human performance.

Chapter 16 addresses the *Regulating PSM and the Impact of Effectiveness* while Chapter 17 looks at a significant area of management which is the *Readying the Organization for Change: Communication and Alignment*. In Chapter 18, case studies are discussed and the *Learning from Incidents* provides great insights into what we continue to learn as an industry. Chapter 19 looks at *Gauging the Effectiveness of Implementation and Measuring the Performance of PSM Activities*, and finally, Chapter 20 is a very reflective discussion on *Human Errors, Organization Culture, and Leadership*. The book closes with an Epilogue of this comprehensive book that covers for the practitioner PSM systems and human factors.

This book offers practitioners a unique powerful, practical and meaningful reference book to lead the integration of human factors with the implementation of PSM systems in organizations.

Finally, the focus of this book is on addressing the human factors in PSM, and with that we hope to bring to the practitioner a very purposeful reference that is enjoyable to read, very useful to learn from, and easy to understand—and most importantly, provides some concepts and ideas to apply.

It has been a pleasure and a wonderful learning journey for us to have worked with such an exceptionally knowledgeable and experienced group of experts and practitioners.

Dr. Waddah S. Ghanem Al Hashmi
BEng(Hons), DipSM, DipEM, MBA, MSc, AFIChemE, FEI, MIoD,
Chief Editor

1

Introduction to process safety management in a practical context

Ahmed Khalil Ebrahim[1] and Waddah S. Ghanem Al Hashmi[2,3]

[1]Bahrain Petroleum Company (Bapco), Kingdom Of Bahrain,
[2]Emirates National Oil Company (ENOC), Dubai, United Arab Emirates,
[3]Energy Institute-Middle East (EI-ME), Dubai, United Arab Emirates

1.1 Prelude

The development of process safety management (PSM) over the past 30 years has been a very interesting and rapid development. In this chapter, we discuss the key drivers in the evolution of PSM, its contribution to the process industry in mitigating serious incidents in addition to low probability and high consequence incidents, and the current challenges faced in effective implementation of the PSM system in the process industry.

The industry cannot ignore the high frequency and perhaps lower consequence events too, but as will be discussed throughout this book, which brings perspectives from some many different practitioners, addressing PSM is a challenging task due to the complexity and diversity of systems and processes.

This chapter will also touch on the concept of human factors and the challenges that arise in PSM. In later chapters, the human factors discussions are explained in depth by other experts who have contributed to this book. The later chapters will thus discuss these human factors and how they relate to various PSM areas of importance based on the working experience of the authors.

Process Safety Management and Human Factors.
DOI: https://doi.org/10.1016/B978-0-12-818109-6.00001-0
1

1.2 Introduction

Occupational injuries and illnesses are of critical concern to governments across the world, because the International Labor Organization (ILO) estimates that every year there are 350,000 deaths due to fatal accidents; almost 2 million deaths are due to fatal work-related diseases and over 313 million workers are involved in occupational accidents causing serious injuries and absences from work.

The ILO also estimates that 160 million cases of nonfatal work-related diseases occur annually. Therefore, it is estimated that every day approximately 6400 people die from occupational accidents or diseases and that 860,000 people are injured on the job [1].

Health and safety concepts and practices have been in existence for at least the last 200 years and it came into existence as a response to the labor movements across the world, especially postindustrial revolution. It was inevitable that workers' health and safety would become an issue as workers suffered fatalities, serious injuries, and ill health because of the workplace activities. It became obvious to the regulators that this is a situation that cannot be tolerated as it would eventually escalate to become a national and a social issue, and that companies must be proactive in finding means of preventing these unacceptable situations.

Hence, the introduction of a legal framework in countries such as the United Kingdom that regulated working hours and conditions, such as the 1833 Factory Inspectorate Act, worker's compensation law in 1884, and later in 1961 the Factories Act, was issued, which consolidated most of the legislation on workplace health, safety, and welfare.

These were followed much later by the infamous 1974—75 Health and Safety at Work Act (HASWA 75), which is a foundational legislation in the United Kingdom and remains one of the most effective and comprehensive single pieces of legislation covering health and safety at work in the world.

Other countries, such as the United States, issued in 1970 the Occupational Safety and Health Act (OSHA) which governed occupational health and safety in the private sector.

Accidents in the workplace can cause significant stress, anxiety, disruption to the business, and the reputational impact on an organization. However, reporting, analyzing, and trending accident information in addition to the establishment of delivery strategy can have a significant impact on an organization, and it could fundamentally change the way people perform their jobs and how decisions are made. This could, in turn, help the organization to increased productivity, employee satisfaction, efficiency, and business continuity.

1.3 The rise of process safety management

The emphasis in the last 100 years in industry has been the prevention of injuries and ill health, which contributed to the improvement of workplace health and safety. Despite the many major incidents that have shocked the world, such as the ones listed later, unfortunately regulators and industry have not given process safety the attention it deserves, until possibly of late, especially at the design and operational stages.

Below are examples of incidents that occurred in the energy sector, which significantly impacted the world in term of human losses, huge cost, and environmental damage, but also influenced the manner process safety is regulated, advocated, practiced, and communicated in industry [2].

Because of the major disasters mentioned in Table 1.1, regulators and industry had no choice but to think of a process and system that would ensure that safety is considered at the inception, design, operation, maintenance stages, and even at the retirement of the facility (cradle to grave).

TABLE 1.1   List of some of some of the major process safety incidents in the last 80 years.

Date/year	Incident	Location	Consequence
October 1957	Windscale fire	United Kingdom	Worst nuclear accident in the UK history.
June 1974	Flixborough (Nypro) explosion	United Kingdom	Explosion totally damaged the plant, destroyed homes and lead to the killing of 28 workers.
March 1979	Three Mile Island accident	United States	Partial nuclear meltdown due to mechanical failures resulted in the release of radioactive gases into the atmosphere.
June1979	Ixtoc I oil spill	United States	Oil spill in which was due to an oil exploratory well blowout, which resulted in one of the largest oil spills.
July 1984	Union Oil refinery	United States	Refinery explosion in killed 19 people. This was in Romeoville, Illinois.
November 1984	San Juanico disaster	San Juanico, Mexico	An explosion at a liquid petroleum gas tank farm killed hundreds and injured thousands. This was one of the worst BLEVE incidents in modern history.

(Continued)

TABLE 1.1   (Continued)

Date/year	Incident	Location	Consequence
December 1984	Bhopal disaster	Bhopal, India	Bhopal gas tragedy in the Union Carbide India Limited (UCIL) pesticide plant in Bhopal. It is the world's worst industrial disaster. Over 500,000 people were exposed to methyl isocyanate (MIC) gas with an estimated immediate death toll of 2,259.
May 1988	Shell Oil refinery explosion	Norco, Louisiana	Hydrocarbon gas escaped and ignited resulted in the killing seven workers and a total of $706 million in damages.
July 1988	Piper Alpha Disaster	North Sea, Scotland, United Kingdom	Platform, North Sea, resulted in the death of 167 workers and losses of $3.4 billion. This incident was rated as the world's worst offshore oil disaster in terms both of lives lost and impact to industry.
March 1989	Exxon Valdez	Alaska, United States	An oil tanker hit the reef, dumping an estimated minimum 250,000 barrels of crude oil into the sea. It is one of the most human-caused environmental disasters.
March 2005	Texas City Refinery	Texas, United States	An explosion occurred at a BP refinery in Texas City, Texas, which was one of the largest refineries in the United States and resulted in the killing of 15 workers and millions of dollars in losses.
December 2005	Hertfordshire Oil Storage Terminal fire	Buncefield, Hemel Hempstead, United Kingdom	In a series of explosions at the Buncefield oil storage depot, described as the largest peacetime explosion in Europe, devastated the terminal and many surrounding properties. Luckily there were no fatalities, but the damage was estimated at £750 + million.
April 2010	Deepwater Horizon Oil Spill—Offshore Production	Gulf of Mexico, United States	Eleven oil platform workers died in an explosion and fire that resulted in a massive oil spill in the Gulf of Mexico, considered the largest offshore spill in US history.

In 1985, the AIChE established the Center for Chemical Process Safety (CCPS), and the first industry standard on PSM was written with involvement of about 20 companies. A few years later API 750: Management of Process Hazards was issued; the purpose of both documents was to prevent the occurrence of catastrophic releases of toxic or explosive materials. It was aimed at dealing and addressing the management of process hazards during design, construction, start-up, operation, inspection, maintenance, and modification of facilities [3]. Much of this development was after the Bhopal incident, which is mentioned in Table 1.1.

In response to industry concern and the government oversight of industries processing and using highly hazardous chemicals and due to the publication of the above two document, OSHA published (29 CFR 1910.119) in 1992 *Process Hazards Management of Highly Hazardous Chemicals*, which defined 14 elements of a process hazards management system.

The elements issued by OSHA through the standard were [4]:

- Process Hazard Analysis
- Operating Procedures
- Process Safety Information
- Training
- Contractors
- Mechanical Integrity
- Hot Work
- Management of Change
- Incident Investigation
- Compliance Audits
- Trade Secrets
- Employee Participation
- Pre-startup Safety Review

Since the introduction of PSM as law in the United States, companies are required to comply with OSHA regulations and mapping it against their internal processes. Some companies had elements of it in existence and had to make sure that it meets or exceeds the legal requirement framework, and hence they commenced conducting gap assessments.

OSHA conducted audits to ensure that companies comply but also to learn how to improve the process and make sure it is practical, does not require retrofitting and major changes to current practices, and give them adequate time for 100% compliance. However, new facilities that were established in the United States after the promulgation had to comply 100% with OSHA PSM requirements.

In time, other companies (and countries) started to appreciate the value of such a system in preventing process incidents and improving the integrity and safety of plants and facilities, and they commenced adopting these requirements, mandated them internally, and also extended them to their contactors. This was also due to major multinationals, many of which had either been headquartered in the United States or otherwise had many facilities operating in the United States, which influenced adoption of similar standards throughout their assets around the world.

Other major companies such as National Oil Companies (NOCs) around the world such as the national companies in the Middle East, Indian subcontinent, and the Far East started also to adopt a PSM system approach rather than simple health and safety management systems.

1.4 Process safety management and human factors

The authors want to emphasize at the very outset of this book that a better understanding of human factors is imperative to understand how the implementation of PSM can be effective. What this book provides are many perspectives on nearly all elements of a PSM system, such as leadership, hazard assessments, management of change, asset integrity, contractor management, and so forth. Each chapter addresses the PSM element, and each author writes about how the human element is significant to understanding each element.

While there has been a growing appreciation of the importance of the human element in accident and loss prevention, this book addresses each of these process safety elements and looks at the human element aspects in a systemic way; given the diversity of the authors and their experience, the following 20 chapters provide great insights to the subject from a pragmatic and practical approach.

Finally, because the book's contributors come with a diverse set of experiences and credentials, the reader will note a significant diversity in the writing styles, yet the focus on the subject and the relevance of both PSM and human factors is clearly provided in each chapter. This book is a practical book, and one that is designed for practitioners to use by providing exceptional insights into different very relevant areas. This book provides, through the experiential knowledge from each author, a unique set of ideas that can be compared and applied to their own working environments.

We have tried to capture the key teachings from each chapter in the closing chapter of this book in a very easy-to-review table.

References

[1] <https://www.ilo.org/global/publications/lang--en/index.htm>. International Labor Organization (ILO), Geneva (accessed 31.10.19.).
[2] The 100 Largest Losses 1978-2017, Large Property Damage Losses in the Hydrocarbon Industry, 25th ed. MARSH Consultants, March 2018. Accessed and downloaded through

website: <https://www.marsh.com/qa/en/insights/research-briefings/100-largest-losses-in-the-hydrocarbon-industry.html>.

[3] Management of Process Hazards. American Petroleum Institute Recommended Practice, API 750, first edition, January 1990.

[4] Process Hazards Management of Highly Hazardous Chemicals, 29 CFR 1910.119, Occupational Safety & Health Administration (OSHA), 1992.

Introduction to human factors and the human element

Ian Randle

C.ErgHF Hu-Tech Human Factors, London, United Kingdom

2.1 Introduction

Human factors (HF) play a fundamental role in process safety management (PSM). There have been significant advances in PSM in recent decades in both the approaches used (e.g., HAZOP, Bowties) and in the use and sophistication of technical safeguards. As a result, hazards are much better controlled, however, major accidents and incidents still happen, and the role of the human in both their prevention and causation is less well understood than the technical safety aspects. HF therefore arguably represents the most important area of focus for future improvements in the broad topic of PSM.

The human is a flexible and adaptable component in the PSM system to be used when it would be too difficult, complex, or costly to use automation or engineered safeguards. A system comprised of both technology and humans offers the greatest scope for both reliability and adaptability, features which any PSM system should possess.

Humans were traditionally seen as the weak link in the PSM chain and the focus of HF activity was for many years on eliminating or mitigating human error, and enhancing human reliability.

However, contemporary thinking acknowledges that the human is an essential component in the system, something that enhances the effectiveness and flexibility of the system and is not just as a source of failure.

To take most advantage from the skills and capabilities humans bring, it is important to make use of established and emerging HF theory and practice.

Process Safety Management and Human Factors.
DOI: https://doi.org/10.1016/B978-0-12-818109-6.00002-2

9

The aim of this chapter is to outline some of the approaches developed to integrate the human into PSM systems and to give examples of some of the practical approaches that have been adopted.

2.2 Terminology and scope

There remains a degree of mystery regarding the scope and application Human Factors (HF) among many in the PSM world. This is caused in part by the array of terms that are used.

To those working as HF professionals and specialists, terms such as HF, human factors engineering (HFE), and ergonomics are often used synonymously and interchangeably. Some might draw subtle distinctions between these terms; for example, using the term "ergonomics" to mean workplace or interface aspects, or the term HF when considering the wider work system, or HFE in relation to projects.

Other terms in use include human systems integration, human and organizational factors, the human element, and human performance. However, there is no single globally used and accepted definition for these terms, they can all be considered part of the same broad discipline, which in this chapter I am just calling "human factors."

A multitude of definitions of HF and the other terms exist, but in its simplest form HF can be described as the design of systems, jobs, and organizations to match human capabilities. The objective is to optimize human and system performance; however, HF is more focused on designing the system to meet the capabilities of humans than to try to change the human to meet the demands of a poorly designed system. The HF approach is therefore distinct from approaches such as behavior-based safety, training and competence, and safety culture, which are covered elsewhere in this book.

The "system" in this context includes the following elements:

- people
- tasks they are performing
- equipment and interfaces they use
- environment in which they are working
- organizations in which they work
- wider culture of the industry and location in the world

What characterizes HF as a distinct scientific discipline is the focus on all elements of the system and their interactions. For example, when considering an operator's response to an alarm, we need to consider:

- characteristics of the operator (training, experience, state of fatigue, etc.)
- design of the alarm interface (colors, sounds, icons)

- task and workload (time pressure, distractions, divided attention, task complexity, procedures)
- environment in the control room (lighting, noise)
- context of the situation and culture in the organization, and all other relevant factors

HF methods and tools have been developed with the system's focus in mind, and it is important when seeking competent HF advice that professionally qualified and certified specialists are used. These include, for example, chartered ergonomists and HF specialists, certified by the Chartered Institute of Ergonomics and Human Factors (UK) and board certified ergonomists or HF specialists certified by the Human Factors and Ergonomics Society (US).

2.3 Why and how human factors are important

The scope of HF within PSM is necessarily broad given the need to consider all relevant aspects of the system. The Health and Safety Executive (HSE) who are the safety regulators in the UK have set out their "top 10" key HF topics (Table 2.1).

Some of these key topics are discussed in this chapter and some in other chapters. A detailed description of all of them is beyond the scope of this book. Nevertheless, this list gives the reader an indication of the scope of HF which an organization will be expected to cover at major hazard sites and provides a good basis for the scope of HF in PSM.

A useful roadmap has been provided by the UK HSE in their Human Factors Operational Delivery Guide on the Control of Major Accident Hazards Regulations (COMAH). This is intended to apply to onshore sites which contain major accident hazards, however the approach is equally applicable to offshore sites (Fig. 2.1).

TABLE 2.1 Key human factors topics: UK Health and Safety Executive.

- Managing human failures
- Staffing
- Fatigue and shiftwork
- Safety critical communications
- HF in design
- Procedures
- Competence
- Organizational change
- Organizational culture
- Maintenance, inspection, and testing

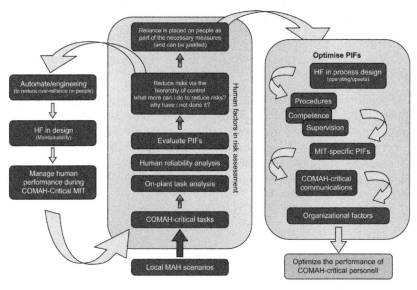

FIGURE 2.1   Human factors roadmap from UK HSE [1].

The roadmap sets out a process for implementing the key HF topics and activities required to control major accident hazards. This or similar processes should form a useful reference source when integrating HF into a comprehensive PSM system.

2.4 Managing human failures

Humans are prone to errors, regardless of how well trained or motivated they are. The use of good-practice HF methods in the design of interfaces, tasks, procedures, work organization, and working environment can all help to reduce the likelihood of errors and should form part of the PSM process. But despite these efforts, errors can never be completely eliminated and will still occur. Thus, the objective should be to ensure that if and when errors do occur that they do not have catastrophic consequences rather than striving to eliminate all errors.

The management of human failure encompasses well-established methods including human reliability analysis (HRA) and human error analysis (HEA). There are a range of structured qualitative and quantitative HRA/HEA methods, which aim to identify error types, causes, and mitigations. It is important to recognize that there are several types of human error that differ in their causes and their preventive measures.

There are several ways to classify errors, with some taxonomies covering numerous categories. For simplicity, I have outlined the classification used by the UK HSE, which identifies four main types: slip (unintended action), lapse (error of omission), mistake (knowledge or rule-based error), and violations (deliberate rule breaking) (Fig. 2.2).

The traditional assumption has been that the appropriate response to human error is to provide further training, this has since found to be ineffective and is now outmoded. Some types of error may be due to inadequate knowledge or training; however, others may be the result of factors such as distraction, fatigue, excessive workload, or poor interface design that are not knowledge-based errors. It is important therefore to understand what factors underlie the error, the so-called performance influencing factors (PIFs) which should be identified before effective prevention or mitigation measures can be devised. It is also important to recognize that accidents and incidents are seldom the result of a single isolated error, there are normally a number of factors involved and therefore detailed HF analysis using the systems approach is required in any investigation. Further information on the various qualitative and quantitative HRA/HEA methods is provided in the references [3–7].

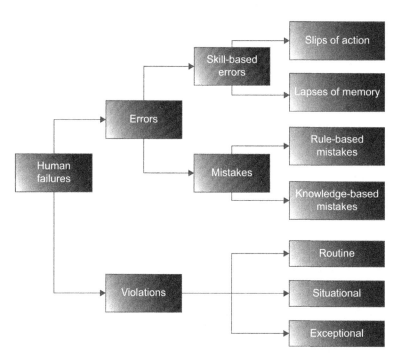

FIGURE 2.2 Types of Human Failure/Error. Source: *HSG48 Reducing Error and Influencing Behavior. HSE. HSE 1999 [2].*

HF approaches to managing human failures (or human reliability) should form an integral part of an organization's PSM system. This is most effectively done by utilizing some of the PSM approaches already in use. For example, Bowtie and HAZOP studies can help to identify where humans form part of the barriers or safeguards to major accident hazards, and to help optimize the use of technical versus human barriers/safeguards as appropriate.

Consider, for example, a typical process deviation identified in a HAZOP: The overpressure of pipework due to a valve being opened in error. Technical (engineered) safeguards to prevent this becoming a major accident might include

- rating the downstream pipework to withstand the overpressure
- fitting pressure relief valves
- installing a high-pressure trip

This is standard good practice in process safety, and these represent effective and error-tolerant safeguards, however they may have drawbacks in terms of added cost, weight, and complexity; increased burden of maintenance and testing; and the potential for loss of production through aberrant trips.

By comparison, human-based safeguards to protect from this process deviation might include

- control access and operation of the valve by permits and procedures
- lock-off the valve to prevent incorrect operation
- provide relevant information and training regarding the operation of this valve.

These measures are simple and inexpensive in comparison to the engineered safeguards, and thus may be attractive particularly in projects and assets where capital expenditure is highly constrained.

Both approaches provide safeguards; however, the lower cost of the human safeguards needs to be balanced against the lower expected reliability of human versus engineered safeguards.

Engineered safeguards need to be carefully specified, procured, installed, inspected, checked, and maintained. This is expensive in comparison to fitting an alarm and developing a procedure for the alarm response.

When considering the reliability of an operator's response to an alarm, this is often overestimated in safety studies such as LOPA where generic failure on demand rates as low as 10^{-5} may be used. Most HF specialists would consider generic human error probabilities to be much higher (more frequent) than this in real-world situations, where 10^{-2} or worse might be more realistic.

It should be recognized that what may be considered the routine task of responding to an alarm includes a number of stages each of which can go wrong:

1. The operator must first perceive the alarm (see it and/or hear it).
2. They must then correctly understand or interpret the meaning of the alarm.
3. Next, they must decide whether any action is required, and if so, what action to take.
4. Finally, they must correctly implement their intended actions.

The human factors that can contribute to each stage going wrong are numerous. These may include inadequate knowledge or understanding of the system, unfamiliarity with the particular scenario, excessive workload or distraction, poor interface design, pressures to maintain production, and so on.

Each of these "failure modes" should be considered in the same systematic manner and with the same rigor as the failure modes for any technical/engineered system, which is designed to monitor and respond to process deviations or upsets.

Engineered systems, particularly those that are safety critical, will have a defined performance standard, which it must meet by design. This performance standard will be maintained by a system of inspection, maintenance, and testing. This is normal practice in the industry.

However, when it comes to human-based safety critical functions (or barriers) the same level of effort is seldom made to ensure that the required level of human performance will be delivered, it is often just assumed that the operator will perform with the necessary accuracy and reliability when required. This represents a significant weakness in the PSM system.

Until recently there was little guidance on how to determine performance standards for human-based barriers. Such guidance has now been provided in the CIEHF White Paper on Human Factors Barrier Management [8]. This sets out six characteristics a human performance standard for barriers should possess:

1. The required human performance should be specific to the threat and the situation.
2. It should be clear who is involved in delivering the required performance. That includes:
   a. Who detects that the barrier function is needed?
   b. Who decides what is to be done?
   c. Who acts to implement the barrier function?
   d. Who is relied on to support the barrier?
3. It should identify the level of competence for all individuals involved.

4. The timing of the performance of the function, including the initiation and time to completion, should be appropriate to the timescale of the threat.
5. The standard for successful performance of the barrier should be defined. For example, criteria could be:
    a. Detect time to trigger the barrier.
    b. Accuracy of interpreting the status of the operation.
    c. Time to initiate an intervention.
    d. Time to complete the intervention.
    e. Maximum number of missed events (i.e., failing to perform the barrier function when it should have been performed).
    f. Maximum number of false alarms (i.e., performing the barrier function when it was not needed).
    g. Tolerance limits for acceptable performance.
6. It should document any expectations made by those who approved the barrier about how operations around the barrier will be conducted that are especially critical to performing its function.

The proposed human performance standard imposes requirements on engineering, design, and organizational arrangements as well as the competence of the individuals involved. This represents a significant step forward in helping to deliver reliable human performance in safety critical situations.

2.5 Safety critical tasks

So far we have been considering the role of humans as a safeguard/barrier to major accident hazards (MAHs). This next section further develops this and provides examples of HF approaches and good practice.

As part of the PSM process it is important to identify, understand, and manage MAH risks. This will include knowing where and when humans are involved in this process. It is standard practice for sites to have a register of safety critical elements and equipment, and a plan in place for ensuring their reliability and effectiveness. However, in my experience it is not yet common practice in PSM to systematically identify and manage the human-based safeguards/barriers. This is an area which requires further attention within the oil/gas and petrochemical industries.

Where a MAH safeguard relies on a human activity (e.g., the response to a critical alarm, or the maintenance of an ESD), then we can consider such activities to be safety critical. The failure of the task or procedure can result in a major accident in the same way as can happen with the failure of safety critical equipment. We call such activities safety critical tasks (SCTs). Each site should have a log or inventory of SCTs and a plan for how they are maintained (see earlier for a discussion on performance standards for SCTs).

SCTs can be identified in several ways, including through HAZOP, Bowties, LOPA, and other safety studies, or from dedicated HF studies. There are also a number of screening tools which seek to identify Safety Critical Operating and Maintenance Procedures. One such tool has been produced by the UK HSE [6].

Once the SCTs have been systematically identified and collated in an SCT inventory then the next stage is SCT analysis (SCTA). This seeks to identify where SCTs can fail and what, if any, additional measures are required to ensure that the procedure can be undertaken as reliably as possible.

There are a few ways to conduct SCTA, but they all have in common the need to understand the task in detail, to list the opportunities for errors with significant consequences, and to identify the factors which might contribute to those errors, called PIFs.

If a detailed written procedure for the task exists then the SCTA process can utilize this to identify the critical steps and errors. The PIFs are best identified by observing or walking through the procedure in the field, so that a full appreciation of the workplace, environment, and equipment factors, which might contribute to an error can be identified.

If no written procedure exists for the task then it will be necessary to undertake a structured task analysis. This involves describing the overall goals of the task, the main steps and relevant sub-steps, and the order and plan for the steps. A common means of recording this task analysis is a structured method known as hierarchical task analysis (HTA). This presents the task in layers, starting with the main steps and showing progressively more detailed steps in layers. The task analysis may not only involve a description of the physical actions of the operator, but also the decisions and cognitive processes which must be undertaken (Fig. 2.3).

If possible, the HTA should be supported by a field observation of the task to ensure all relevant details, and PIFs are captured.

The purpose of the procedure review and task analysis is to identify where and when critical errors can be made. These are then subjected to human error or reliability analysis (HEA/HRA) to understand the type or types of error which may be involved. There are many ways to undertake a HEA/HRA, ranging from simple informal qualitative methods through to detailed structured quantitative methods. A description of the different types can be found in the HSE publication HSG48 [3] and the textbook by Kirwan [5].

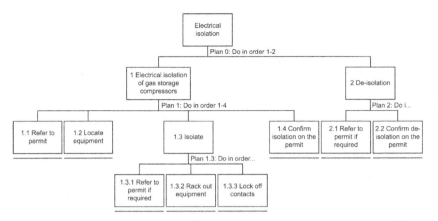

FIGURE 2.3   Example of a hierarchical task analysis (for electrical isolation).

Typically, a qualitative approach is adequate and yields sufficient information to devise additional practical measures to reduce risk. Quantitative HRA methods can be useful for input into QRA studies but are time-consuming to conduct and are reliant on generic human error probabilities which may not be applicable to a particular situation.

As discussed earlier it is important to differentiate between the different types of error so that the potential causes and best means to eliminate or mitigate them can be identified. Any such measures should be discussed with the end-users of the procedure or those performing the critical task to ensure that they are satisfied that the measures will be usable and effective in practice.

SCTA should be undertaken by an experienced HF Specialist with input from operations, process and engineering. Further information on SCTA can be found in a guide from the Energy Institute [7].

2.5.1 Identifying safety critical tasks from HAZOP studies

In a capital project opportunities exist to identify and eliminate SCTs from the early stages. In the FEED-stage HAZOP the focus is on identifying where humans are sources of process deviations (e.g., valve opened in error) and, also where they are involved in the safeguard to the deviation (e.g., response to an alarm, executing a procedure).

This early identification provides opportunities to design the human out of the MAH safeguards, and to identify where the safeguards and system will benefit from human intervention. Doing this early in the project gives maximum scope for designing the asset and systems to enhance human performance where it is needed, or to replace with engineered systems where it is not.

The HAZOP also provides information on where humans are sources of critical process deviations. This can be used to guide HF studies to ensure that appropriate effort is put into the design of specific interfaces, equipment, signage, and work environments to minimize the likelihood of such errors occurring in the first place.

In later stages of a project the opportunities to design-out humans will be fewer, but more detail of the equipment and process will be known, and therefore it will be possible to undertake HF assessments such as SCTA and HEA.

Where it has not been possible or desirable to eliminate the human from the system (e.g., during start-up where parts of the technical safeguards are necessarily suppressed or disabled, and therefore manual monitoring and control maybe required) then the focus of HF attention will be on ensuring human performance and reliability during safety critical procedures.

These critical procedures should be developed and assessed using good HF principles, and subject to SCTA and HEA as described above. Guidance on the development of safety critical procedures is available from the UK HSE and the Energy Institute [6,7]. From experience, it is possible to identify straightforward and practical steps which significantly reduce error opportunities and consequences in critical procedures.

2.6 Human factors in design

Building in good HF design to an asset from the outset represents one of the cost-effective means to ensure safety as well as reliability and operability. The design of equipment, workplaces and human-machine interfaces (HMIs) should make it easy for the operator to do it right and difficult for them to do it wrong. The ideal time to do this is during a project design and specification phase.

HF input to capital projects originally started as the application of HF standards which were integrated into the engineering specifications. These standards comprised largely of generic guidelines on accessibility, equipment layout and interface design. However, this approach lacks the fundamental aspects of the HF 'systems approach' to design. That is, it does not take account of the specific target audience, the organizational factors, the particular tasks being undertaken, or the range of working environments encountered for the asset being designed.

The practice of HF integration in oil and gas projects has a relatively short history. It is only within the past ten years or so that the major oil/gas operating companies have started to embrace the systems approach to HF, something which has been embedded in other sectors for much longer.

HF should be applied to all key phases of major projects: from early concept design, through front end engineering design (FEED), to the detailed engineering, procurement, and construction (EPC), hook-up and commissioning, and early operate phases.

Industry guidance on HF in major projects has been published by the International Association of Oil and Gas Producers (IOGP) [9]. This sets out a roadmap for HF integration and identifies the key HF activities in the various phases of a project. Importantly, this includes both work-place design requirements and the human contributions to major accident risks, which sets it apart from previous industry guides that focused mainly on the former. The IOGP guide also sets out roles and responsibilities and competency requirements for HF assistance on a project. This guide is becoming the standard upon which the major operating companies are basing their internal HF standards and processes (Fig. 2.4).

One area of design which is particularly important to PSM is the central control room (CCR). This is where many of the SCTs will take place and often where incidents and major accidents will be monitored and managed. It is important, therefore, when designing and assessing a CCR and the control and safety systems within it that HF best practice is used.

The main international standard for the ergonomic design of control rooms is ISO 11064 [10]. This sets out a process for the design and development of the CCR and associated rooms to ensure that is fit for purpose in the given context, it does not attempt to provide 'standard' layouts.

Another useful reference source for assessing safety critical functions in control rooms is CRIOP: A scenario method for Crisis Intervention and Operability analysis [11]. CRIOP differs from traditional CCR assessments in that it introduces a series of scenarios, including upset and emergency conditions, and systematically assesses the ability of the control center to handle the situation by working through the scenarios with relevant stakeholders.

Thus, if seeking to ensure that the CCR and its operators will function as required during upsets and emergencies, then the CRIOP approach provides a more robust verification than traditional assessments. It therefore represents a useful tool in PSM.

Equally as important as the design and layout of the control room is the design of the HMI for the process control, fire and gas, and alarm systems. Key sources of best practice in HMI design are ISO 9241 Ergonomic Requirements for Office Work with Visual Display Terminals [12] and NUREG-0700 Human-System Interface Design Review Guidelines [13]. These detailed standards form the basis for many internal Company HMI Specifications. Further good practice on HMI design is provided in

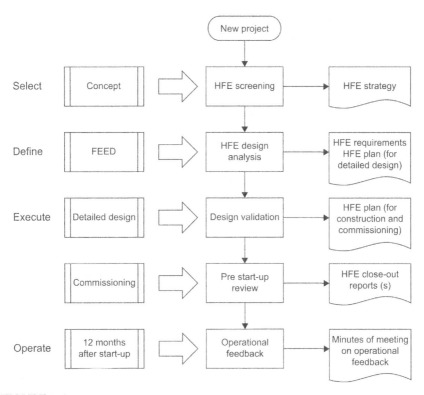

FIGURE 2.4   Human factors in the project lifecycle—from IOGP 454 [9].

EEUMA 201: Process plant control desks utilizing human-computer interfaces—a guide to design, operational, and human interface issues [14]. EEUMA 191 [15] is the industry standard for alarm design and management, providing useful guidelines on the number, rate and presentation of alarms, together with rules for prioritization and suppression to aid human performance in alarm handling.

2.6.1 Future directions for human factor in design

One future direction for HF relates to the continuous drive within the industry to reduce manning for safety and cost reasons. Replacing people with technology (e.g., enhanced instrumentation, Artificial intelligence and augmented reality) is an aspiration for many new projects and existing assets, but the implications are not always properly considered. For example, a skilled process technician who walks the plant daily on an inspection round/tour may develop tacit knowledge on how to keep the plant running smoothly, learning the idiosyncrasies of the plant based on their "on the ground" experience. If the technician is removed and the daily walk-rounds are replaced by increased instrumentation, and a more complex control system in a remote-control room, there is a risk that the skilled functions of the technician are not adequately replaced by the technology and the same level of efficiency and plant availability will not be achieved.

There is a role here for HF to analyze the skilled behaviors and information acquired by the technicians on the ground, and help identify how these can be effectively replicated by the control and instrumentation system. However, this level of HF influence on a project is exceptional. The challenge for HF is to become sufficiently embedded in project processes to allow this type of input to become routine.

2.7 Conclusions

HF is a very broad-based discipline drawing on human psychology, physiology, anthropometry, industrial design, engineering, architecture, and user interface design. It seeks to optimize the performance of systems in which humans are a significant part, and therefore to enhance their safety, operability, and efficiency.

This chapter has set out to provide a brief introduction to the scope of HF and how this relates to PSM. It has not been possible to cover all relevant areas and therefore the focus has been on managing human failures, SCTs, and designing error-tolerant, user-friendly work environments. The approaches and methods described are based on good HF practice and the author's practical experiences in the implementing of HF in high-hazard industries around the world over the past two decades.

A comprehensive list of Standards, Codes, Guidance and Regulations relevant to Human Factors and Ergonomics in Process Safety is given in Appendix 2.

References

[1] Inspecting Human Factors at COMAH Establishments (Operational Delivery Guide) UK HSE. v1 2016.
[2] Reducing Error and Influencing Behaviour. HSG 48. HSE 1999.
[3] J. Reason, Human Error, Cambridge University Press, 1991.
[4] S. Dekker, A Field Guide to Understanding Human Error, Routledge, 2014.
[5] B. Kirwan, A Guide to Practical Human Reliability Assessment Paperback, CRC Press, 1994.
[6] HSE OTO 1999 / 092 - Offshore Technology Report - Human Factors Assessment of Safety Critical Tasks.
[7] Guidance on Human Factors Safety Critical Task Analysis. Energy Institute, 2020.

[8] Human Factors in Barrier Management. A White Paper by the Chartered Institute of Ergonomics & Human Factors, 2016.

[9] Human Factors Engineering in Projects. Report No. 454. International Association of Oil & Gas Producers, 2011.

[10] ISO 11064 - Ergonomic Design of Control Centres.

[11] CRIOP: A scenario for Crisis Intervention and Operability Analysis. SINTEF, 2011.

[12] ISO 9241 Ergonomic Requirements for Office Work with Visual Display Terminals.

[13] NUREG-0700. Human-System Interface Design Review Guidelines, 2002.

[14] EEUMA 201: Process Plant Control Desks Utilizing Human-Computer Interfaces - A Guide to Design, Operational and Human Interface Issues, 2019.

[15] EEMUA 191: Alarm Systems: A Guide to Design, Management and Procurement, 2013.

Leadership and process safety management

Waddah S. Ghanem Al Hashmi[1,2]

[1]Emirates National Oil Company (ENOC), Dubai, United Arab Emirates,
[2]Energy Institute-Middle East (EI-ME), Dubai, United Arab Emirates

3.1 Introduction

It is significant for organizations and their practitioners to understand the fundamental role of leadership in general in organizations and how it links with process safety management (PSM) leadership and corporate governance. In this chapter, some aspects of leadership with respect to driving behaviors will be discussed, but more importantly this chapter will discuss how leadership and leadership failures can lead to very significant incidents and loss.

How management can influence human behavior and human factors is something which is becoming so important to understand to prevent the trajectory of incidents. Using a simple model of the before and after the incident and how management can assess the relationship between efforts exerted proactively and their relationship is process safety performance is extremely important.

The role of leadership in general in organizations and how it links with PSM leadership and governance within the context of the organization's strategy. Leadership models will be discussed, but more importantly in this chapter will discuss several case studies and give comparative review of where leadership failures led to very significant incidents and

Process Safety Management and Human Factors.
DOI: https://doi.org/10.1016/B978-0-12-818109-6.00003-4

loss. How management can influence human behavior and human factors (this will be discussed in this chapter).

The roles that leaders can play in helping with the success and effectiveness of PSM will also be discussed.

3.2 Process safety elements: leadership

Over the past few decades various PSM models and management systems have been developed, such as OSHA in the United States, or the Energy Institute (EI) in the United Kingdom, as well as company-based systems. Some have since transformed into more operational excellence and operational integrity management systems. One very common and cardinal element is leadership. All these systems call for the PSM to be managed. In general, leadership must give three key things: (1) the overall vision to achieve the strategy, (2) the steer needed to achieve the strategy, and (3) the compelling reason to motivate others to achieve the strategy.

We find that leadership must provide that motivation to the organization, to drive it toward achieving PSM excellence or what is commonly known as operational excellence.

So, what are the key components of these elements and what are the expectations driving?

When developing any framework, including PSM systems or otherwise, leadership is one of the four fundamentals along with risk management, continuous improvement and implementation. In fact, all the elements such as commitment and accountability, policies, and monitoring/reporting/learning and assurance, to name a few require a maturity and effectiveness in the four fundamentals. (See OGP Report No. 510, June 2014.)

More recent work in leadership, and especially safety leadership, suggests that leaders should now go beyond commitment to stewardship and advocacy. This means that they provide that compelling drive toward safety excellence while ensuring that effective support through the provision of resources and giving authority and empowerment.

Leadership drives the safety culture within the organization and this drives the behavior of people within the organization (IOGP 452). But leadership needs to be accountable, and safety is everyone's responsibility and not just the responsibility of the safety or EHS functions and the managers only.

A good safety culture is one where an organization is able to develop and promote the values and beliefs of the organization the overtime shifts where safety becomes a core value of the organization and an integral part of operations.

International oil and gas producers (IOGP) make reference to the work of T.R. Krause (2005), which talks about the seven key safety leadership characteristics and associated behaviors that can influence safety culture:

1. *Credibility*: what leaders say is consistent with what they do.
2. *Action Orientation*: leaders act to address unsafe conditions.
3. *Vision*: leaders paint a picture for safety excellence within the organization.
4. *Accountability*: leaders ensure employees take accountability for safety-critical activities.
5. *Communication*: the way leaders communicate about safety creates and maintains the Safety Culture of the organization.
6. *Collaboration*: leaders who encourage active employee participation in resolving safety issues promote employee ownership of those issues.
7. *Feedback and recognition*: recognition that is soon, certain, and positive encourages safe behavior.

So how does leadership apply to the process safety elements? In PSM systems, the focus is on an integrated set of elements all with expectations and that need to work toghther. It is the operational discipline, which is exercised by the organization that makes the PSM systems effective, and the value proposition is presented through the effective discipline which is employed within an organization.

Leaders have to seek assurance that the safety management systems working are effective; they should have insights about system effectiveness, and they use this to better understand performance. This comes through periodic reviews and the engagement with the key perfromance measures and data. An understanding of the key result areas is fundamental to better understanding of key performance indicators (KPIs) as measures of assurance. The main issue, from experience, with very senior leadership in organizations and those who are far removed from the operational workings, is that they cannot see the relationships and the causalty between certain failing barriers and a safety event such as a high potential near miss or an incident.

What we shall try to illustrate through three examples of where leadership's insights could help in the steer and prevention of major incidents or otherwise provision of preventative measures, and so on.

So let us look at some examples.

3.2.1 Example 1: Contractor safety management

One of the critical elements of effective contractor safety managmenet as will be discussed more extensively in Chapter 13, Contractor Management, by Lutchman and Akula is the drive toward managing the organization's expectations with any contractor at the tendering stages where the scopes are made clear and the companies' evaluation of the technical and safety capabilities is driven by the responsible sense of effective risk managament. The commercial pressures are significant on all projects, in large and small organizations. Leaders that ensure that an effective, fair, and risk-based approach to the selection, evaluation, award, and management of contractors is of paramount importance.

When organizations award contracts to subsafety standard contractors, they create a great risk, which now needs to be managed more closely by the front-line supervisors. A noneffective evaluation sends a message that safety culture is something which the management wants to promote, but not at any expense! This can drive a significantly negative behavior. In turn supervisors, while continuing to try their best to manage the contractors, will always feel that the leaders of the organization who should understand that awarding of a contract to a substandard contractor creates risks, which are what the policies and company safety mantra tries to aviod.

In this case the leaders of the organization who set the tone and should steward the safety policies, should not only be aware and empahtehic to these issues, they need to be greatly supportive to helping the supervisors/managers implement the safety standards with their contractors with no compromise. Getting the right level of contractor in by setting an appropriate acceptablity criteria to start with is the best way to manage such issues.

3.2.2 Example 2: Prestartup safety reviews

One of the biggest pressures on plant managers and project managers of new-build plants or those who have gone through a shutdown cycle is the startup phase. This is probably the most critical and dangerous time in a plant's operational cycle as will be discussed further in Chapter 12, Prestart Up and Shutdown Safety Reviews. The basic and most important tenant of any operational management system, and in fact operational excellence, is that operational discipline, which is practiced within an organization. During the preparations for shutdowns and startups, planning managing risks through careful assessments and working very closely with all the parties involved in shutdown and startup are the key to a safe and successful shutdown and startup.

The preactivity safety reviews are very critical. The reason for this is ultimately that the plant is never designed to operate generally under the startup and shutdown condition, therefore the plant behavior is sometimes difficult to estimate or predict. Using a similar approach to a managmenet of change process, as discussed extensively in Chapter 9, Management of Change, by Ram Goyal, leadership must provide the right level of support and must give oversight to the planning activities.

From experience, delays in startups can cost organizations greatly. With proper planning and solid leadership support, effective plans executed can make sure this can be avoided.

From personal experience in a refinery, by creating a sector-wise leadership organization with centralized reporting supported by an effective plan with daily reporting and progress coordination, a full 100,000 + barrels per stream day refinery can go through a full turnaround and inspection shutdown less than 30 days and with no LTIs. This was achieved through a cascaded delegation through a coordinated leadership setup with a very effective centralized planning function.

Leaders effectively set the standard of performance which is expected and create the drive toward the key perfromance areas based on the key result areas. These help establish the KPIs, which are monitored. Leaders focus on the key result areas and the main goals of the startup being achieved rather than only look at the KPIs and remain contempt with the quantitative performance and seek to understand better the more qualitative aspects of the shutdown and startup.

It must be appreciated that leaders must understand their impact on the workforce, the management, supervisory, and operational levels. These work under a different dynamic mode during shutdowns, turnaround and inspection, and startups. They generally work longer hours, around the clock, and with no day-off for 30−40 days. The level of leadership support and emphathy needed to help ensure the goals are achieved is of paramount importance, especially that with increased chance of fatigue, greater stress, and multicommunications breakdown with several organizations (e.g., multiple shutdown contractors and inspectors) the chance for human error and reduced human performance is significantly increased.

3.2.3 Example 3: Asset integrity management

Asset integrity management (AIM) as discussed in Chapter 8, Asset and Mechanical Integrity Management, by Hirak Dutta, is based on three key areas that include technical, operational, and design integrity management. At the heart of AIM is the leadership and human resource capacity and capability. The aspects relating to competencey and capacity are discussed in Chapter 8, Asset and Mechanical Integrity Management, along with the other aspects of AIM, however, what is very significant is the leadership aspects of ensuring the effective implementation of AIM.

There are several significant failures that relate to asset integrity deficiencies: by the inherently unsafer design, the operational integrity management failures, or otherwise due to the ineffective management of change and technical integrity issues. Leadership's involvement in understanding the asset's performance over time is crtically important to the maximum extraction of value from an asset. Managing it effectively requires leading the design, technical, and operational excellence of an asset.

But leadership's role goes beyond only managining the three aspects of AIM. Leadership's role is to drive the right behaviors by facilitating the right competencies in the workforce, oversight of planning, challenging the design integrity, and making sure that the discipline in the management of change processes is maintained.

With AIM, the human factors are a very imprtant aspect to understand. Processes relating for example to maintenance must be effective in as far as they facilitate rather than create barriers, which make it a cumbersome task to maintain and operate an asset. Behavioral drivers may require consideration and with critical tasks, layers of protection by ways of increased assurance reviews may be required.

Finally, assets maintained to effectively operate over extended periods of time with a greater reliability in many plants and operations is a matter of pride for many engineers and operators. Leadership must therfore inspire the workforce to creating a high reliability asset by ways of working collaboratively, with a high degree of pride in their craftsmanship and creating key performance indicators that should drive the right positive behaviors. To this end, leaders must be involved in helping facilitate dialogue with the workforce and empower them by investing in their development and by giving them the appropriate level of autonomy while creating the right balance between creativity in effective troubleshooting and the right level of discipline in assurance.

So, from the earlier three examples, one can see that the role of leadership is critical to the actual effective implementation of PSM system elements. The same would be applicable to any of the PSM elements as leadership actually creates the drive and inertia needed for the implementation of any system.

3.3 Understanding the leadership challenges

The development of the PSM systems have been for many years driven as a regulatory compliance requirement, especially in North America or otherwise have been very much driven from best practices through the development of industry. Process safety has been a subject to many developments in the past 30 years, and improvements in the elements have come about from learnings from major incidents such as the Flixborough disaster in the United Kingdom in 1974 and the Piper Alfa disaster in 1988 and many others in the past decade or more.

All these incidents, and after effective root-cause analysis, led to the development of these systems through a better understanding of the main failures and failure modes and in that the main components of the elements covering various PSM elements where developed. So, there has been some great benefits gained from learning from incidents.

However, one of the key drivers behind writing this book and bringing in such a diverse set of authors with diverse experiences, knowledge, and discourses was mainly to understand why when, for almost four decades of the existence of process safety and PSM systems, is the industry still having significant incidents which are leading to loss of life, massive losses in assets, and resources. Even more so today, in an interconnected world of supply chains, such incidents have dramatic impacts reaching far beyond the perimeters and vicinities of the assets where the incidents occur.

The challenges facing leadership are clearly seated in understanding the human factors behind many of the ineffective or erroneous actions (or otherwise inaction or delayed response) that lead to the incidents and their escalation. Can training be the solution to reducing human error? If that were the case, then we would have done very well in eliminating almost all incidents. In the chapters to follow, we address some of these key issues. In fact, in the next chapter, which addresses "the awareness of risk, complacency and the normalization of deviance," a good introduction into this subject is made.

Leaders need to understand the human element inasmuch as they must understand the fundamentals of PSM, or in fact the fundamentals of business operations, finance, marketing, and general management. They must understand better the management of human factors with respect to the drivers behind human behavior inasmuch as understanding the fundamental inherent vulnerabilities and the creation of safer environments with reduced errors and incidents through design, operating procedures, and competency development.

3.4 Process safety leadership: a model

Integrated management systems today and many of the (ISO) management system standards developed for quality management, environmental management, and occupational health and safety management are built on the Deming cycle, which has become known as the Plan-Do-Check-Act (PDCA) cycle.

Using this and looking at the PSM management system, a simple model that looks at three aspects that leadership needs to drive the effective implementation. In this leadership needs to have qualities in addition to the technical and behavioral competencies. A high-level understanding of governance, risk management, and control is very important and a good working knowledge of the policy development, implementation, and the role of leaders in

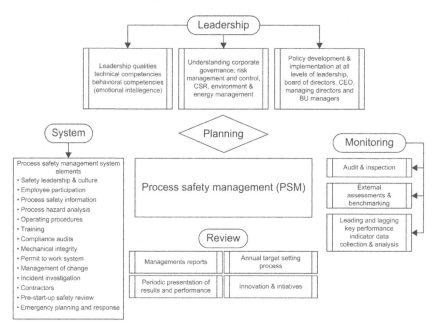

FIGURE 3.1   PSM leadership model.

different levels with the organization is very important. Below shows a suggested model for a PSM leadership in Fig. 3.1.

3.5 Monitoring and managing process safety management performance

Understanding leading and lagging indicators is critical for managing performance. Leaders must understand at the very minimum:

- Leading indicators are considered predicative indicators.
- Leading indicators can be used to evaluate the weaknesses that can be corrected to reduce occurrences.
- Lagging indicators tend to be outcome oriented—describing events already occurred—but also have value as they indicate events that may reoccur if the problems/root-causes are not addressed.
- They are both equally important and they give important and insightful perspectives.
- They are one of the most significant tools that managers and leaders use—giving a "snapshot"—understanding what the values mean is very critical.
- Highly dependent on the target-setting methodology and process.

The most comprehensive code that addresses PSM KPI has been for the past decade the *US API Code of Practice* 754:2010, which gives four tiers for the PSM performance:

- *Tier 1*: Loss of Primary Containment (LOPC): Events of Greater Consequence;
- *Tier 2*: LOPC Events of Lower Consequence: Process Safety Events that can provide good learning for organizations;
- *Tier 3*: Challenges for Safety Systems: that is, the consequence is stopped short of T1 and T2 and indicators at this level provide additional opportunity to identify weakness and make corrections;
- *Tier 4*: Operating Discipline and Management System Performance Indicators: typically looking at closure of audit and incident findings, trainings and drills completed, safety-critical equipment inspections, and so on.

But when developing KPIs, in the energy sector, it is important to create an indicator list with the right definitions and definitions need to be set in line with industry norms to create the right level of benchmarking. This is to ensure that performance is comparable to the industry at large and creates the drive toward betterment through setting improvement targets. This must be driven by the leaders on the organization.

It is recommended, therefore, that key performance measures are made:

- *Reliable*—unbiased scale, specific and discreet
- *Repeatable*—similar conditions will yield similar results
- *Consistent*—units and definitions are consistent throughout Group
- *Independent of external influences*—not influenced by pressure factors
- *Relevant*—to the operation—translate in to informed action
- *Comparable*—comparable with other similar indicators
- *Meaningful*—sufficient data exists to measure positive/negative changes
- *Appropriate for the Intended Audience*—aggregated, normalized, reporting period data
- *Timely*—information is useful for that time-period
- *Easy to Use*—indicators should be easy to measure not too complicated
- *Auditable*—for assurance on the accuracy of the information.

Finally, the use of KPIs is a very good and effective way of monitoring performance like using a dashboard. It is, however, extremely important for leaders to appreciate the qualitative aspects of the performance such as the causality between drivers, operating conditions, human factors, and the PSM performance in the organization. As such, understanding the context is and will always be where leaders can better create the basis for a high reliability organization rather than a compliance-based culture where the KPI drives the performance.

All the KPIs should reflect on the actual key performance measures, the value drivers, and therefore when leaders look at the numbers, a sense of what the performance indicators are telling leaders individually and collectively in the context of the greater external and internal factors/aspects are very important. As such information should reflect on the impacts on and from the plant performance, incidents, shutdowns and turnarounds, extreme weather conditions, projects ongoing in the asset and even macro-economic conditions etc. is very important.

3.6 Way forward and chapter concluding remarks

In this short chapter, we have addressed the key literature regarding the safety leadership characteristics and associated behaviors that can influence safety culture. Our focus was less on the developing safety cultures in as much as trying to pragmatically talk about the role of leadership in driving the effective implementation of PSM systems.

Some examples have been given with respect to three of the PSM elements to illustrate how leaders must drive the elements toward effective implementation. The chapter also addressed the challenges of leaders who need to understand human factors and the human elements as they much understand all other technical and business-related aspects of their roles in the organization.

A simple model for PSM is presented. This chapter concluded with some pragmatic recommendations on monitoring and managing PSM performance through the balanced approach of leading and lagging indicators.

4

The awareness of risk, complacency, and the normalization of deviance

Ronald W. McLeod

Independent Human Factors Consultant, Ron McLeod Ltd.

4.1 Introduction

In this chapter, a profoundly important aspect of safety management is discussed and related to human behavior and tendencies to circumvent the very standards that, for example, a process safety management (PSM) system provides. Adherence and compliance to following systems designed to protect people and assets must be reinforced, and a sense of mindfulness, or "chronic unease" in both the leadership and working workforce must exist.

In this chapter, theawareness of risks, underlying complacency and the normalization of deviance are discussed with simple, very purposeful, and illustrative examples.

4.2 Toward understanding deviation

A tendency to complacency about risk in the face of repeated success, and the normalization of deviance that is closely associated with it, can be extremely difficult to overcome: it appears to be deeply embedded in the human psyche.

Many organizations in safety critical industries go to great lengths to try to avoid complacency taking hold. This is reflected in the level of interest in topics such as "chronic unease" [1], and "safety mindfulness" [2,3], among others. Avoiding complacency is fundamental to attempts

FIGURE 4.1   My bike opposite target café.

to set the goal of "zero" accidents and serious incidents as an expectation rather than a target [4].

I took the photograph on Fig. 4.1 on my way home after a day's cycling. My bike is at the bottom left. On the right-hand side is a café where I had planned to stop for coffee.

While cycling, I had been thinking about writing this chapter: How could I convey what I wanted to about the challenge of ensuring people are aware of risk at the moments they are about to make decisions and act? And what practical action can be taken at the operational level to work toward a culture that actively challenges complacency and is effective in preventing it? As I reached the café, I stopped, got off my bike and started to cross the road. Then I stopped as I realized what I was doing.

The area where I was about to cross the road has a history of accidents, including fatalities. Drivers regularly park—even double park—at the side of the road and cross the road to access the shops. At certain times of the day, this is a very busy road. I had many times seen, and even commented on, drivers who would stop their car and rush to cross the busy road in a manner that showed seemingly no awareness either of the danger they were in or the problems they were creating for other road users. Not surprisingly, numerous accidents and fatalities had occurred here in the time I had been living in the area. I knew that.

Several changes had been made to reduce the risk of accidents: traffic had been reduced to a single lane, a parking bay had been built, and traffic lights had been installed I knew all of this. And yet, despite that knowledge, what had I done? I had stopped my bike on the roadside about 30 m short of the parking bays and was about to step onto the road to cross to the café. This was to avoid cycling some 70 m further to the traffic lights that had been provided as a safe crossing point. Why had I done that? Why, despite my knowing that the area had been associated with fatal accidents, had I mindlessly ignored the features that the local council had provided specifically to make the area safe?

Reflecting on this example of my own unsafe behavior, I recognized three features that seemed to capture what I had done:

1. First, I knew the area was hazardous. Had I been asked beforehand, I could have explained that there had been a number of serious incidents, including fatalities, at the location. I knew the risk.
2. Second, I knew the right thing to do: The safe way of accessing the cafe was to cross at the traffic lights.
3. Third, I chose to do something else. My knowledge of the hazards and risks, and of the safe way of behaving, was not sufficient to cause me to go to the small effort to behave in the safe way expected by the council who had provided features to protect my own safety.

I realized that I was behaving mindlessly. And I was motivated by the very thing that I had been thinkingabout and reflecting on as I was cycling for the previous hour: I had been complacent about the risk.

4.3 What does complacency mean for process safety?

Reflecting on my behavior while cycling, and recognizing the three elements behind how I had acted, I wanted to understand more about the nature of complacency, and what others in the scientific, research and applied communities thought about it. It is a complicated and confusing topic that overlaps with concepts such as mindlessness, normalization of deviance, and routine violations. Does it mean anything?

In its common usage, complacency can be thought of as the opposite of a sensitivity to risk: the opposite to a sense of safety mindfulness or chronic unease. It is also seen as overconfidence born from repeated success [5]. In their classic book *Managing the Unexpected*, describing the characteristics of high reliability organizations that has become central to so much thinking in safety management, Weick and Sutcliffe noted that "if people assume that success demonstrates competence, they are more likely to drift into complacency, inattention and predictable routines. What they don't realize is that complacency increases the likelihood that unexpected events will go undetected and accumulate into bigger problems" [2, p. 56].

A reading of the technical and scientific literature suggests at least three ways of thinking about complacency:

- organizational
- automation-induced
- situational

4.3.1 Organizational complacency

One of the main conclusions reached by the Report to the President into the Deepwater Horizon tragedy was that the incident occurred as a consequence of a "culture of complacency" that existed throughout the organization: "Absent major crises, and given the remarkable financial returns available from deep-water reserves, the business culture succumbed to a false sense of security. The Deepwater Horizon disaster exhibits the costs of a culture of complacency" [6, p. 11].

This kind of complacency exists at the organizational level: complacency that has become part of the social structure of the organization rather than being identified with any specific individuals. As the Sociologist Diane Vaughan puts it: "mistakes are systematic and socially organized, built into the nature of professions, organizations, cultures and structures" (, 7, p. 415).

In a discussion of organizational complacency in NASA following the loss of the shuttle Columbia in 2003, Denning [8] identified several contributing factors behind organizational complacency: experts who are overconfident in their own abilities; an over-reliance on success as signs that the organization is properly controlling risk; group-think; and excessive confidence in technology. Denning suggested a range of strategies for avoiding complacency at the organizational level such as giving those with deep technical knowledge independence from those with management and operational responsibility, and introducing formal processes that allow professionals to express unpopular points of view—and that ensures they will be listened to.

4.3.2 Automation-induced complacency

There is a sizable body of research dealing with what is termed "automation-induced complacency" (e.g., [9, 10]). This kind of complacency is associated with people whose role is to function as "supervisory controllers" of highly automated systems, such as modern commercial and military aircraft and process control systems. In supervisory control, the role of the human operator is largely to monitor what the automation is doing, and to be ready and able to intervene if and when needed, either in response to prompts and system alarms, or when people detect or suspect that the system is not working properly.

Automation-induced complacency arises when the human has become overreliant on the automation: they trust it too much. The operator fails to monitor sources showing how the automation is behaving as often as they should. Automation-induced complacency is essentially a suboptimal allocation of the operators' attention, arising from unrealistic—indeed, overoptimistic—beliefs about how reliable the automation actually is.

4.3.3 Situational complacency

Situational complacency is the subject of this chapter; the kind of complacency I experienced on my cycle ride. It is the complacency that leads people to make decisions and take actions in a specific time and place that are not consistent with the immediate risks, but that allows them to avoiding going to effort, or that make sense in the situation they are facing at the time. Situational complacency is similar to Professor James Reason's concept of routine violations: "...if the quickest and most convenient path between two task-related points involves transgressing an apparently trivial and rarely sanctioned procedure, then it will be violated routinely"[11, p. 196].

Situational complacency refers to decisions or actions that colleagues and peers in the organization, if they knew about them, would not sanction or approve of. Often it involves decisions that are intended to be temporary or are intended as a one-off "quick fix" to a short-term problem: "normal service" is expected to be resumed after a short time. Table 4.1 contains some examples that illustrate situational complacency.

4.4 Complacency and the normalization of deviance

Complacency as intentional deviation from expected safe ways of working as a result of repeated success, has similarities with the concept of normalization of deviance (NoD). NoD came to prominence in connection with the investigation into the loss of the space shuttle Challenger in 1986: "No fundamental decision was made at NASA to do evil: rather, a series of seemingly harmless decisions were made that incrementally moved the space agency toward a catastrophic outcome ... a process that can create a change-resistant worldview that neutralizes deviant events, making them acceptable and nondeviant" [7, p. 410].

TABLE 4.1   Examples of situational complacency.

Operations managers who allow operations to continue while knowing

- safety systems are not in service, or inspections are are out of date;
- high hazard product streams are aligned to atmosphere because treatment systems are not available;
- high hazard systems are being operated beyond their safety limits;
- levels of overtime are being worked beyond company limits;
- contractors are authorized to work in conditions with known health and safety problems;
- workarounds or incorrect equipment is being used without adequate management of change.

Team leaders who do any of the following, or who know colleagues are doing them but do not intervene

- issue or close a permit-to-work without visiting the work site;
- carry out hazardous tasks without using the correct protective equipment;
- re-use gaskets in high hazard systems due to lack of stock;
- do not comply with lock-out/tag-out (LoTo) procedures;
- continue operating for a long time knowing an alarm has been bypassed;
- improvise tools to help operate manual valves or loosen bolts;
- apply isolations without ensuring proper traceability.

Complacency does not need to become normalized: it can be a one-off event—indeed, it may have led to the deviation that became normalized in the first place. It does, though, require the intention not to do what is known to be expected. Whereas the original deviance can be unintentional (a slip, lapse, or mistake) or a purely technical failure, it only becomes normalized when people know about the deviance and allow it to continue.

There is now broad awareness of NoD in most sectors that manage operational risk. Banja [12] studied why healthcare professionals fail to comply with procedures established as basic standards of care (such as not washing hands or using sterile equipment). He identified the same kind of contributing factors as other industries: from a belief that rules are unnecessary, stupid or inefficient, or a belief that "they do not apply to me," to workers being afraid to speak up.

Preventing normalization requires that deviances are recognized for what they are as soon as possible after they occur. And taking action immediately to acknowledge that the decision or action was deviant, to correct it, and to understand what motivated the deviance in order to prevent the likelihood of its recurrence. The key strategy to preventing a deviance from becoming normalized lies in not allowing it to go "unattended, unappreciated or unresolved for an extended period of time" [12, p. 139].

4.5 The elements of Situational Complacency

With reference to my cycling example presented at the beginning of this chapter, recall the three elements of complacency:

1. Knowing the risks;
2. Knowing what is expected as the safe way of behaving; and
3. Choosingto do something different.

These three elements are summarized in Fig. 4.2, along with some examples of the key features associated with each element.

Considering the second and third elements first, there are many reasons why operators might not be aware of what is expected of them to work safely: from lack of competence and unclear responsibilities, to poorly documented, ambiguous, or inaccessible procedures. Sometimes the organization itself is not clear on what the safe way of working is. In healthcare for example, there are frequently no clearly documented processes or procedures for many treatment paths. Similarly, there are many reasons why people choose to deviate: from lack of understanding of why procedures are needed or lack of confidence in them, or a belief that procedures do not apply to them, to a perception that there is not enough time to follow the procedure, or that complying will interfere with production goals. These are complex topics.

The remainder of this chapter will focus on the contribution to complacency from the first of the three elements: knowing the risks. Essentially, this boils down to the difference between knowledge and awareness.

Awareness means "the ability to directly know and perceive, to feel, or to be cognizant of events … the state of being conscious of something" [13]. An "awareness of risk" means an understanding of the potential adverse consequences of an activity that is effective in ensuring decisions, actions and the ways people behave and interact with other people at the worksite are consistent with ensuring risks are properly controlled. It is not about simply identifying risk, or giving people

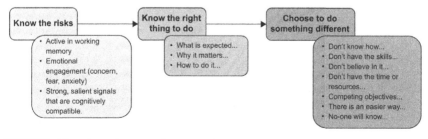

FIGURE 4.2   Three elements of situational complacency.

the knowledge such that, if asked, they could tell you what the risks are and how they are expected to be controlled.

Awareness of risk must bring a level of emotional engagement or connection with personal values: some sense of anxiety, worry, or concern, or even a degree of fear about what could go wrong, together with a personal motivation and commitment to behaving safely ("I never cross a road on a red light"). At the very least, it means a conscious awareness that something could go seriously wrong *right now* if you do not behave safely. It is this emotional engagement and personal commitment—though without affecting the individual's ability to think clearly and perform to a high standard—that is so important to developing a sense of chronic unease and safety mindfulness.

Arstad and Aven [14]emphasize the central place that awareness of the signs and indicators of risk has in situational complacency: "the necessary evidence was knowable, recognizable, understandable, it should have been weighted properly and acted upon adequately" [14, p. 116]. Complacency relies on the fact that although the individual "knows" the risks, the signs or signals indicating the presence of the danger are not sufficiently salient or powerful to influence decisions or behavior: they do not impact on the "cognitive now" [15, p. 31].

4.6 David's story

Consider the following story. Although it is fictitious, it contains elements of decisions and choices that should be familiar to most parents, if not most adults.

> David is 10 years old. At the weekends, he likes to ride his bike on the street with his friends. One day, David tells his father that the brakes on his bike are not working properly. His father inspects the bike and finds that the brakes are badly worn. He also notices that the straps on David's helmet are broken. He tells David to stop cycling until he can fix the brakes and buy him a new helmet.

> The next Saturday, David again asks his father if he can play on his bike. His father has forgotten to fix the brakes or buy a helmet, but is too busy to fix them that day. He ties a piece of string to the helmet to act as a strap and tells David he can play, but to be careful.

> David enjoys his day playing with his friends. On Sunday, he again asks his father if he can play on his bike. His father says yes: he decides the brakes will last for another day and the string on the helmet seems to be working OK.

What are the hazards and risks in this story: What is the worst that could happen? What controls was David's father relying on to protect his son, and which of them did he know were not working properly? What deviances did he know about? Was he complacent? Had he normalized the risk? Most importantly, why would David's father allow his son to cycle knowing that two of the controls he was relying on were not working properly?

David's father must have believed that tying a piece of string to the helmet and telling David to be careful will provide adequate protection. If he did not believe that, he would either not have allowed his son to cycle, or he would have fixed the problems he knew about. The key question is, in the moments when he decided that it was safe for David to cycle, what did David's father believe the risk to be? What did he think was the worst that could happen?

Returning to David's story, here is one of two alternate scenarios for what happened next.

Scenario A:

> The following Saturday, David again asks his father if he can play on his bike in the street. David's father has again forgotten to fix the bike or to buy a new helmet. He again says yes David can play on his bike, and again tells him to be careful.

> While playing with his friends, David cannot stop his bike, and runs into a neighbors' new car, scratching the paintwork. David fell off his bike during the accident and cut his leg. The neighbor says David's father must pay for the damage. David's father is very angry: he tells David he was careless and bans him from riding his bike for a month.

David's father's assessment of the risks is the same as the previous weekend. And he knows that the issues in the safety defenses still exist. But he is prepared to continue to rely on the same controls—the string and relying on David to be careful. Expecting David to be careful is not unreasonable in itself. But it is unreasonable when it is an alternative to having properly functioning brakes. Not only is his decision deviant, it is well on the way to becoming normalized: nothing bad happened on the previous weekend, so he relies on them again.

Here is an alternative scenario for the same weekend.

Scenario B:

> While playing with his friends, David cannot stop his bike. He runs into a neighbors' new car and falls off his bike. His helmet falls off and another car hits him on the head when he is on the ground.

> David dies.

Were the hazards and risks any different between scenarios A and B? No. Did David's father imagine that his son could be killed? Surely he cannot have imagined such a dreadful outcome: if he had, he would have acted differently.

Among other things, David's story illustrates one of the biggest challenges in any risk assessment activity, or any attempt to raise awareness of risk: having the "requisite imagination" [1] to appreciate, to visualize, and to take seriously, what might actually happen if control of a hazard or hazardous operation is lost. When faced with a job site, or any other situation, as human beings we find it very difficult to visualize what might go wrong unless we have direct personal experience or other reason to make a strong mental association with adverse outcomes.

There is an extensive scientific knowledge base about how people develop awareness of risk and how that awareness impacts on thinking, decision-making and behavior. Among the most relevant is what we know about differences between the two styles of thinking referred to as "System 1" ("Fast") and "System 2" ("Slow") [16]. Psychologists know a great deal about the many sources of irrationality and bias that can be associated with System 1 thinking [15,17–20].

Despite this extensive knowledge base, the expectations industry holds about the ability of people to identify and maintain awareness of risk at a work site can be fundamentally in conflict with how the human brain thinks and makes decisions, especially in the presence of uncertainty. References 15, 17,18 and 19 include many examples of how the biases and thought mechanisms associated with "Fast" thinking can lead to failure to identify and act on risk in the workplace.

As an example of unrealistic assumptions about the ability of people to assess risk, it is common practice to require Job Risk Assessments, or an equivalent, to be carried out before starting any hazardous work. One company also requires what they refer to as a "Dynamic Risk Assessment" to be conducted during the performance of a job.

"Dynamic Risk Assessment" is defined as a "...continuous, intentional process of assessing risk to identify and mitigate hazards/aspects to control an expected outcome. This can be cognitive or verbal." The term "Cognitive Risk Assessment", is further defined as: "The basic level of practical thinking and judgement that individuals use to assess and mitigate hazards/aspects (i.e. walking down stairs)." Once work on a job has started, this company therefore expects that "employees shall be intentional in continually assessing the risk/impact to control the expected safe outcome of each job."

The expectation that, on top of performing physically and cognitively demanding work, workers can be "intentional" in "continuously assessing" risk is unrealistic. It is also inconsistent with a very large body of knowledge about how humans think, and how we perceive and assess risk in real time: It is simply not how the human brain works. Human beings are not capable of consciously paying continuous attention to any details of their environment or task performance for more than relatively short periods of time. Indeed, if the kind of expectations behind the "Dymanic Risk Assessment" process were somehow to be enforced, the demands on cognitive processes such as attention and working memory would be such that the people involved would be incapable of performing even moderately complex tasks.

4.7 Conclusions

Ultimately, the purpose of all human factors (HF) activities in a PSM program is to ensure people know what they are expected to do to ensure risks of major accidents are contained; that they understand and believe in the importance of complying with the expected ways of working; that they have the competence, motivation and ability to be able to comply; and that the design and layout of work systems—both equipment as well as work controls—allows them to do what is expected to the standard needed.

In this chapter, we are focused on HF solutions and strategies that can be implemented by leaders who are close to the operational front-line where situational complacency has its greatest influence. While PSM systems provide an excellent framework, operational discipline informed and supported by a concerned and situationally aware leadership is of paramount importance.

The awareness of risk, and the reasons for complacency, are complex and deeply psychological. But what does it mean in practice for those tasked with managing HF in a process safety program for operational assets who do not have a background in psychology or the time or opportunity to apply the knowledge? There are several key points:

1. Being realistic about the complexity of how people think and make judgments about risk. And in particular, being aware of the many biases and sources of irrationality associated with our "fast" thinking brain that can lead people to jump to conclusions and make decisions that ignore the available facts. And using that realism and awareness to challenge processes and procedures that make assumptions about what people can or will do that are unrealistic or inconsistent with human nature.

   Most importantly, leaders must recognize the simple fact that human beings have a deep psychological drive to avoid going to the effort to engage our slow thinking brains, to think things through and make decisions and judgments based on the evidence in front of us: we are hard-wired to find the easy way to do things. This truth has been central to much research and applied work by psychologists and HF specialists for decades (see Ref. [16] for a summary of the evidence). Despite this, innumerable processes and procedures assume and expect that people will go to effort and follow procedures, despite there being an easier way. The rallying call for many in HF for many years has been that work systems—including equipment as well as processes and procedures—should be designed such that *the easy way is the right way*. If that principle were taken more seriously, the incidence of complacency would likely be dramatically reduced.

2. Recognizing that there is a big difference between simply telling people about risk—whether that is in formal training, in documented procedures, or in the signage and warnings and other signs placed around hazardous equipment—and that knowledge actually being effective in influencing thinking and decision-making in real time. For example, see my discussion of local rationality in connection with the explosion and fatalities at the Formosa Chemicals plant in 2005 [15].

3. Encouraging people and teams to be realistic, rather than optimistic, in visualizing what could go wrong if they don't comply with expected safe work practices. That means encouraging a mindset that does not assume things will be OK because nothing has gone wrong before. And challenging those who do not believe an outcome is credible because they cannot imagine it happening, or have never seen or heard of it themselves. Or, worse, those who are not prepared to acknowledge worst cases as being credible because they are inconvenient, or create difficulties in getting the job done.

4. Making the effort to ensure that the indicators of risk—the signs and signals intended to remind the individual in the workplace of the presence of hazards and to make them aware of the risk if work is not properly controlled—are powerful enough (sufficiently "salient") to influence decision-making and judgment in real time at the workplace. Applying good practice in Human Factors Engineering (HFE) in the design of the work environment and interfaces to hazardous equipment has an important role to play in making signs of risk salient. Good HFE in the design of work systems can go a long way to avoiding "weak" signals of danger (chapter 8 of Ref. [15], discusses the role of HFE in making "weak" signals "strong" in the design of work systems).

What difference could these points have made to my own complacency during my cycling expedition illustrated in Fig. 4.1? All of the features put in place to make the area safer were passive: safe parking bays, traffic lights, bollards, and a reduced road width. None of them could actively increase my real-time awareness of the risk, or stimulate my thought processes in a way that would motivate me to go to the effort to behave safely.

There are however innumerable examples of thinking that has gone into the design of roads and highways around the world that appear to be effective in influencing awareness and thought processes at the time and in the situation where increased risk exists.[1] There is much the process industries could learn about improving the awareness of risk at the frontline by a close study of research, experience, and thinking by those responsible for the design of road transport systems.

Solutions and strategies to combat organizational complacency, strategies that will be effective in ensuring the unjustified self-confidence that goes along with complacent thinking does not permeate an organization's values, beliefs, and become part of how the organization thinks about and manages risk, lies largely—it certainly starts—with senior leadership.

Solutions to automation-induced complacency have to lie primarily—strategies for countering them are primarily to do—with the design of automation. In particular, the allocation of functionality between people and automation when automated systems are being conceived, and the design of human machine interfaces that support people in the role of supervisory controllers in increasingly highly automated systems.

The solution to deviances arising from situational complacency becoming normalized lies with those working closest to the situation where the deviance occurs: often that means at the operational frontline. The solution comprises three components:

1. Being prepared to recognize and name deviance for what it is. A significant first step would be simply to use the word "deviance" when somebody knows that a decision or course of action is not consistent with what the company wants or expects. Call it for what it is: it is "deviant."
2. Leadership being willing to listen and acknowledge deviance, and to act quickly to stop it in its tracks.
3. A continuous effort of individual transformation toward safety mindfulness (built on elements such as self-reflection, personal commitment, and teaching others).

Perhaps the most important lesson anyone needs to learn, is that as soon as a deviation gains a foothold, the step toward its becoming normalized is a small one: "quick fixes" not being put right at the first opportunity; David's father allowing his son to continue to play on his bike on the second weekend. Once a deviation becomes normalized, it becomes an issue of organizational complacency. Which is much harder to fix.

---

1 Space in this chapter does allow a discussion of these examples, though chapter 20 from ref 13, . includes an example.

4.8 Final thoughts

To conclude, here is an example taken from NASA's Aviation Safety Reporting System [21], that illustrates the kind of behavior associated with a culture that does not tolerate deviances.

> This report is to highlight my concern about personnel who are not active working crew members on a flight (jumpseaters) but take it upon themselves to arm and disarm aircraft doors. I have personally had this situation happen, and I have witnessed it happening to fellow working crew members.

> My intent . . . is to bring to the attention of the company . . . an action that should be discouraged and discontinued due to its ability to impact the safety and security of an armed aircraft door . . . I think a note or bulletin needs to be sent out to each and every flight attendant explaining proper procedures so that complacency does not breed an opportunity for a fatal outcome.

Developing a culture that does not tolerate deviances is, of course, a challenge for most organizations that face the pressures and commercial realities of running hazardous operations in competitive markets. Unfortunately, the only way to prevent deviances arising from complacency turning into normal practice is to build a culture that simply does not tolerate deviance. That, essentially, is what many companies have been working toward through implementation of programmes such as the IOGP's Life Saving Rules [22] and approaches to establishing Golden Rules for Process Safety [23].

References

[1] L.S. Fruhen, R.H. Flin, R.W. McLeod, Chronic unease for safety in managers: a conceptualisation, J. Risk Res. 17 (8) (2013) 969–979. Available from: https://doi.org/10.1080/13669877.2013.822924.

[2] K.E. Weick, K.M. Sutcliffe, Managing the Unexpected., second ed., Jossey-Bass, San Francisco, 2007.

[3] L. Mazur, L.B. Marks, R.W. McLeod, W. Karkowski, P. Mosaly, G. Tracton, et al., Promoting safety mindfulness: recommendations for the design and use of simulation-based training in radiation, Adv. Radiat. Oncol. 3 (2) (2018) 197–204. Available from: https://doi.org/10.1016/j.adro.2018.01.002.

[4] SPE, Getting to Zero and Beyond: The Path Forward, Society of Petroleum Engineers, 2018.

[5] J. Bell, N. Healey, The Causes of Major Hazard Incidents and How to Improve Risk Control and Health and safety management: a review of the existing literature. Report HSL/2006/117. UK Health and Safety Laboratory, 2006.

[6] National Commission on the BP Deepwater Horizon Oil Spill and Offshore Drilling, Deep Water: The Gulf Oil Disaster and the Future of Offshore Drilling – Report to the President, National Commission, 2011.

[7] D. Vaughan, The Challenger Launch Decision: Risky Technology, Culture and Deviance at NASA, second ed., University of Chicago Press, 2016.

[8] S. Denning, Challenging Complacency. ASK Magazine. April. National Aeronautical and Space Administration, 2006. Available from: <https://appel.nasa.gov/2006/04/01/challenging-complacency/>.

[9] S.M. Merritt, A. Ako-Brew, W.J. Bryant, Λ. Staley, M. McKenna, A. Leone, et al., Automation-induced Complacency potential: development and validation of a new scale, Front. Psychol. (2019). Available from: https://doi.org/10.3389/fpsyg.2019.00225.

[10] N. Moray, T. Inagaki, Attention and complacency, Theoretical Issues in Ergonomics, Science 1 (4) (2007) 354—365.

[11] J. Reason, Human Error, Cambridge University Press, Cambridge, 1990.

[12] J. Banja, The normalization of deviance in healthcare delivery, Bus. Horiz. 53 (2010) 139—148.

[13] Wikipedia contributors, Awareness. Wikipedia, The Free Encyclopedia. Retrieved from: <https://en.wikipedia.org/w/index.php?title = Awareness&oldid = 923127693>, 2019 (accessed 28.10.19).

[14] I. Årstad, T. Aven, Managing major accident risk: Concerns about complacency and complexity in practice, Safety Science 91 (2017) 114—121.

[15] R.W. McLeod, Designing for Human Reliability: Human Factors Engineering for the Oil, Gas and Process Industries, Gulf Publishing, Oxford, 2015.

[16] D. Kahneman, Thinking, Fast and Slow, Allen Lane, London, 2011.

[17] McLeod, R.W., The impact of styles of thinking and cognitive bias on how people assess risk and make real-world decisions in oil and gas operations. Oil and Gas Facilities. October. Society of Petroleum Engineers, 2016.

[18] R.W. McLeod, Human Factors in Barrier Management: hard truths and challenges, Process. Saf. Environ. Prot. 110 (August) (2017) 31—42. Available from: https://doi.org/10.1016/j.psep.2017.01.012.

[19] R.W. McLeod, From Reason and Rasmussen to Kahneman and Thaler: styles of thinking and human reliability in high hazard industries, in: R.L. Boring (Ed.), Advances in Human Error, Reliability, Resilience and Performance, Springer Books, 2018, pp. 107—119.

[20] IOGP, Cognitive Issues Associated with Process Safety and Environmental Incidents. Report 460. International Association of Oil and Gas Producers, 2012. Available from: <https://www.iogp.org/bookstore/product/cognitive-issues-associated-with-process-safety-and-environmental-incidents/>.

[21] NASA., No place for complacency Callback: Aviation Safety reporting System. March. National Aeronautics and Space Administration, 2017. Available from: <https://asrs.arc.nasa.gov/publications/callback/cb_446.html>.

[22] IOGP, IOGP Life Saving Rules. Report 459. International Association of Oil and Gas Producers, 2019. Available from: <https://www.iogp.org/life-savingrules/>.

[23] CCPS, Project 289: Golden Rules for Process Safety for specific technologies, Center for Chemical Process Safety, 2018.

5

Competence assurance and organizational learning

Janette Edmonds
The Keil Centre Limited, Edinburgh, United Kingdom

5.1 Introduction

Within Process Safety Management (PSM), erroneous assumptions can be made that people will perform perfectly every time, and it is not uncommon for "competence" to be cited as a risk control or safeguard in safety analyses. However, people do and will continue to fail to perform as expected, and if "competence" is claimed as one of the safeguards, particularly against a major accident, then it needs to be properly substantiated.

Assuring human performance is a fundamental component of effective process safety assurance. Developing the required competence for job tasks, particularly those that are safety critical, contributes to achieving the performance required of the humans within the system. Competence itself is a continuum which is developed over time and initiated through training, but it must translate into safe behaviors if safety is to become a day-to-day reality.

However, it is not just about having competent individuals to operate and maintain the process safely, but also about maintaining a safe culture and evolving safer practices through effective organizational learning.

This chapter discusses the need to assure human performance as part of PSM, which includes processes to assure competence. It also discusses how organizational learning and safety culture can strengthen the claim that the system is safe.

Process Safety Management and Human Factors.
DOI: https://doi.org/10.1016/B978-0-12-818109-6.00005-8

5.2 Assuring human performance

Human failure does not happen by accident but is largely driven by the tasks people are asked to do, the tools and equipment they use, and the environmental and organizational context within which they work. It therefore follows that if all these aspects are designed well, then the potential for human failure is reduced.

It is necessary to gain an understanding of the human within the system and how people can contribute to major accidents. The assurance of human performance is explained later on in this book in Chapter 15, Human factors and PSM compliance assurance activities. This describes the different types of human failures, the conditions that make failure more likely, and the techniques and management processes that can be implemented to identify and reduce human failure vulnerabilities, both during design and in operation. A key method described is Safety Critical Task Analysis (SCTA). Within this technique, a list of Safety Critical Tasks (SCTs), where human failure could initiate, escalate, or fail to prevent a major accident, is derived. Each SCT is analyzed in more detail to identify how human failures could occur, the potential consequences of the failure, and how risks are intended to be controlled. Where the risk controls are inadequate, additional controls are identified. This was also discussed in Chapter 2, Introduction to human factors and the human element.

The risk controls aim to prevent the human failure in the first place (through task and engineering design). If it cannot be fully prevented, then there is a need to enable detection and recovery from the failure. If it cannot be recovered, then engineering or human/ organizational controls need to be implemented to provide the final safety net.

Just as for engineered controls, human and organizational controls must work for them to be of any value. These might include, for example, having well-designed procedures that can be followed without confusion (not just the fact that there is a procedure), ensuring a manageable workload so that critical information is not missed, using effective communication protocols so that instructions are not misunderstood, and good fatigue management programs to avoid performance decrement when making important decisions.

Effective training and competence assurance and a good safety culture are two examples of human and organizational controls. If they are well managed, they can provide an important safeguard against major accidents and process safety events. If they are poorly managed, they can help to instigate them. An example where a lack of competency contributed to major accident is presented in Box 5.1.

An example where poor safety culture contributed to a major accident is presented in Box 5.2.

BOX 5.1

Lack of Competence Contributes to Catastrophe

In September 1998, an explosion occurred on the Esso Gas Plant in Longford, Australia, killing two people and injuring eight others. It also led to a shutdown of the gas supply for the whole of the state of Victoria, leaving homes and businesses without gas, hot water and heating for 20 days. The event was particularly devastating for industries and commercial enterprises, especially those in the hospitality trade which were reliant on the gas supply.

The plant is the onshore receiving terminal for oil and natural gas production from offshore platforms in the Bass Strait. It has three gas processing plants and one crude oil stabilization plant. On the day of the accident, a pump supplying heated lean oil to a heat exchanger in Gas Plant 1 went offline for four hours. The normal operating range of the heat exchanger was 60°C–230°C, but due to the pump failure, it cooled to an estimated −48°C. The pumping of hot lean oil was resumed to thaw the ice that had formed on the unit, but the temperature of the oil being pumped into the heat exchanger was at 230°C. This caused the heat exchanger to rupture due to cold metal embrittlement. The hot oil pump failure was only one aspect of the accident process, but one of the key conclusions was that operators did not fully understand the dangers of cold metal embrittlement and were ill-equipped to deal with the events on the day of the accident. There were several other human factors issues, including poor management of organizational change, inadequate procedures, supervision and safety culture and poor plant design.

(Longford Royal Commission [1])

5.3 Competence and human performance

Competence is defined as

[T]he ability to undertake responsibilities and to perform activities to a recognized performance standard on a regular basis. It involves a combination of practical and thinking skills, experience and knowledge, and may include a willingness to undertake work activities in accordance with agreed standards, rules and procedures. *(Office of Rail Regulation (ORR) [3]).*

BOX 5.2

Poor Safety Culture Undermines Safety [2]

In April 1986, the No. 4 reactor at the Chernobyl Nuclear Power Plant near Pripyat, Ukraine, exploded, killing 30 people within the first few months. The explosion and resulting fire released large quantities of radioactive particles into the atmosphere. The longer-term effects, including cancer, are predicted to be responsible for many more deaths. It is considered to be the worst nuclear disaster in history.

At the time of the accident a safety test on the nuclear reactor was being performed which included a simulation of an electrical power outage. The purpose was to support the development of a procedure for maintaining cooling water circulation using residual energy from the momentum in the main turbine generators as they spun down to cover the time required for the emergency generators to start up. There was a potential safety problem that could cause the nuclear reactor core to overheat and the design engineers had failed to identify a suitable solution on three previous occasions.

The fourth attempt was delayed by ten hours, which meant the operating shift that had been prepared was not present. The test supervisor also failed to follow the procedure, leading to an unstable operating condition. In addition, several safety systems were intentionally disabled and, combined with inherent design flaws, this led to an uncontrolled nuclear reaction. The operator's knowledge of the design flaw was also lacking (although the positive void coefficient in the fuel rod design was pretty much unknown to anyone working in nuclear safety until after the incident). A large amount of energy was suddenly released, causing the superheated cooling water to vaporize and the reactor core to rupture. An open-air reactor core fire then released the airborne radioactive contamination for around 9 days.

It is possible to lose competence as well as gain it, as shown in Fig. 5.1 (based on [3]). The dashed arrow represents a loss of competence. Training and development fit into the first stage of the continuum, and on-the-job development, supervision and reassessment fit into the second stage.

Clearly, there is a need to keep personnel at the point of being unconsciously competent and avoid the person returning to the unconsciously incompetent stage. In addition, there is a need to develop and maintain competence as new tasks are introduced to the remit of the operator. In any organization, there is likely to be a mix of competence levels, so

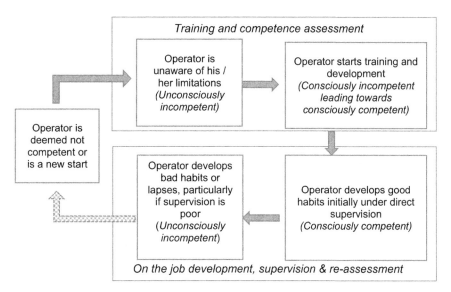

FIGURE 5.1 The competence development continuum.

there will be some more competent than others and the situation does not necessarily remain static over time.

Training is typically the means by which a basic level of capability is developed. It can take many forms, such as classroom-based training, computer-based training, instruction, reading, simulation training, and so on. Training alone should not be regarded as sufficient for developing competence; on-the-job development, refresher training, coaching, and mentoring are also significant aspects of competence development.

Some tasks that personnel perform will be individual, and some tasks will be team tasks. Regardless of task type, competence is comprised of knowledge, skills, and attitudes as follows:

- The *underpinning knowledge* of the job includes the production processes and chemical reactions, the engineering systems, work processes and operating procedures.
- *Technical* competencies are task-related skills and enable the job to be functionally completed, such as an electrician being able to wire up a junction box. Technical competencies are typically job or occupation specific and the objective is to meet minimum standards of performance.
- *Nontechnical* competencies are attitudes and behaviors that contribute to excellent performance and require knowledge and skill during the conduct of the role; examples include the ability to schedule tasks effectively, work well in a team, coach, or motivate others, and communicate effectively.

All three aspects are necessary for effective and safe performance: knowledge, technical competencies and nontechnical competencies. Nontechnical competencies are equally important as technical competencies.

Competence needs to be effectively managed using a comprehensive Competence Management System (CMS). This provides the vehicle for developing and assuring the competence of its personnel.

The Office of Rail Regulation (ORR) developed a CMS for safety critical work on the UK railway system [3], and this has been adopted by the UK Health and Safety Regulator as best practice for the chemical and process industries. The CMS is a five-phase cycle, as summarized in Fig. 5.2.

5.3.1 Phase 1: define the requirement

The competency requirement involves identifying tasks that could affect safety. The starting point is the SCTA (described earlier), to understand which tasks are undertaken, to decide which are safety critical, to review how they are conducted and to identify the vulnerabilities to human failure. Safety assessments, such as hazard and operability (HAZOP) studies, should also be reviewed as part of this process to identify any stated dependence on competence.

Once the competency requirement is identified, it is then necessary to either select the appropriate competence standards (where one already exists) or develop new standards of competence to define the performance

FIGURE 5.2   Five phases of the competence management.

and knowledge requirements for the safe performance of tasks. A competence standard should reflect the type of competence required, including the correct way to perform the task and the underpinning skills, knowledge and attitudes/ behaviors.

Although specific qualifications may be part of the competence standard, there may be a need to customize the standards of competence to include all modes of operation and scenarios that are relevant to the site.

The competence standard provides a testable description of competence with measurable criteria for judging performance and the level of performance evidence required.

5.3.2 Phase 2: design the CMS

There is a need to develop the procedures and work instructions for operating the CMS itself to ensure the CMS is consistently applied and achieves the intended results.

It is also necessary to establish how each competence standard is to be met and assessed. A comprehensive guide to the development and assessment of competence standards is provided for the hazardous industries (HSE, [4]).

The methods of recruitment, training, and assessment for staff and contractors will then need to be devised. The training, development, and assessment requirement will need to be established, noting that the training required for new recruits may differ from that for existing staff and contractors.

5.3.3 Phase 3: implement the CMS

This phase is about implementing the CMS, including selection and recruitment, training and development plans, training delivery timetable, and assessment against the relevant competence standards.

It is important to ensure that personnel only perform the work if they have adequately demonstrated they are competent to do so (supervised or unsupervised as relevant). There may be a need for additional training and development, or if this is unsuccessful, then redeployment or another performance management strategy may need to be taken.

5.3.4 Phase 4: maintain competence

This phase is about monitoring and reassessing personnel through planned and unplanned observations to ensure that competence is being maintained.

Any organizational change, introduction of new technology, or a change to the work context needs to be accompanied by an update in staff competence in relation to that change, if safe performance is to be maintained.

Any substandard performance is dealt with during phase four. This requires a method for improving competence and a procedure for removing those who consistently fail to meet the competence standard. However, it is also important to understand the reasons for any perceived lack of competence as it could be due to factors outside of the person's control. These might include aspects of the job, for example, a change of equipment; attitudes within the organization, for instance behavior driven by the work culture; or the individual themselves, for example, their attitude to work, whether they have an injury, whether age is a factor or whether there are personal issues affecting their behavior.

Finally, record keeping in relation to competence is essential and this can be used to plan further training and development in an organized and systematic way.

5.3.5 Phase 5: audit and review the CMS

The final stage in the cycle is to audit and review the CMS to ensure it maintains integrity. This includes checking the quality of competence assessments (including the competence of those operating and managing the CMS) and ensuring the assessment process remains effective. Any deficiencies or deterioration in the CMS should be rectified and reassessed.

5.4 Wider organizational learning

The previous section concentrated on developing "competent" individuals and teams to operate and maintain a process safely. However, organizational learning must go beyond this level in the organization.

> [Organizational learning involves embedding lessons in the organization itself, not in the individuals who make up that organization. *(Hopkins 2012 [5])*

Andrew Hopkins is an Emeritus Professor of Sociology at the Australian National University and has been actively involved in analyzing major accidents, including the explosion at Esso Longford, the Texas City refinery explosion in 2005, and the Deepwater Horizon blowout in 2010. One of his key areas of focus has been related to the failure of organizations to learn from incidents and accidents. Despite the devastation caused by major accidents and incidents, such events provide a golden opportunity to learn and prevent them happening again in the future. His work provides some insight into why this does not always happen, and what organizations can do to improve corporate memory.

Effective organizational learning requires:

- A comprehensive understanding of the human factors relevant to an incident (or indeed to potential incidents). It is insufficient to stop the analysis at the realization of human failure. There is a need to understand why someone has, or could, fail.
- Implementing organizational change to embed the learning and lessons within organizational practice. This may include changes to the organizational structure, its procedures and processes, its resourcing, its priorities, and its performance indicators. This is much more than the typical top down sharing of lessons learned.
- Leadership to take an active role to drive the implementation of organizational change. Senior leaders need to recognize the need for, and the value in, securing these changes.
- Personnel to be informed in an active and engaging way. The process industry has typically taken a passive and often technical approach to implement learning, such as using email digests, warnings and flyers. However, for deep learning to occur, there needs to be reflection and action. Lardner and Robertson [6] developed a scenario-based practical learning approach for learning from incidents using actual case studies. Their recommendation was to make learning and training more active and engaging generally, not just for the dissemination of learning from incidents.

A review of organizational learning was undertaken by the Keil Center on behalf of a client in the nuclear sector [7]. One of the conclusions arising was that effective organizational learning is a key component of a safe culture. Learning is limited by what individuals can identify, the knowledge and experience base to learn from, and by the culture around them which can facilitate or hinder the valuing and subsequent embedding of that learning. Conversely effective organizational learning is supported by cultural values of openness and excellence; learning mechanisms that encourage information flow, challenging assumptions, and aiding systems thinking; and commitment of resources. Both for organizational learning and effective safety culture, it must be driven from the top of the organization.

5.5 Concluding remarks

For PSM to be effective, there needs to be assurance of the human performance of those operating and maintaining the system. This requires a comprehensive understanding of human failure vulnerabilities and the implementation of controls and safeguards to reduce the risk.

Claims are often made about the competence of personnel within safety analyses, and indeed the human performance assurance process, but if "competence" can truly be relied upon, there needs to be an effective means of developing and maintaining it. This requires a structured mechanism for managing and assuring competence.

However, it is insufficient to rely solely on individuals within the system without understanding the necessity for the organization to learn and embed structures and practices that will survive beyond the personnel at a single given point in time. This is a key component of a safe culture, and a safe culture is required to support organizational learning.

References

[1] Dawson, Sir Daryl Michael and Brooks, Brian John. The Esso Longford Gas Plant Accident. Report of the Longford Royal Commission. Government Printer for the State of Victoria, 1999.
[2] Wikipedia, free online dictionary. Chernobyl Disaster. Online dictionary accessed August 2019. (https://en.wikipedia.org/wiki/Chernobyl_disaster).
[3] Office of Rail Regulation (ORR), Developing and Maintaining Staff Competence, second ed., Railway Safety Publication 1, 2007. HSE, ISBN 0 7176 1732.
[4] Health and Safety Executive. Competence Assessment for the Hazardous Industries. Research Report RR086, 2003.
[5] Andrew Hopkins, Disastrous Decisions: The Human and Organisational Causes of the Gulf of Mexico Blowout, CCH, Australia, 2012.
[6] R. Lardner, I. Robertson. Towards a Deeper Level of Learning from Incidents: Use of Scenarios, IChemE Hazards XXII Conference, paper 54, 2011.
[7] J. Wilkinson, H. Rycraft. Improving Organisational Learning: Why Don't We Learn Effectively from Incidents and Other Sources? IChemE Hazards XXIV Conference, paper 159, 2014.

6

Integration of human factors in hazard identification and risk assessment

Hari Kumar Polavarapu

EHS Director, Emirates National Oil Company (ENOC), LLC, United Arab Emirates

6.1 Introduction

While the evolution of the chemical industry since time immemorial was driven by human curiosity, and endeavors to learn new applications for various resources available in their natural settings, the modern processing industry owes its genesis to the frenzy of innovations necessitated by the industrial revolution. The discovery of oil and its refining technology pushed the boundaries further with the introduction of a multitude of petrochemicals that today are integral to our daily lives.

This chapter will address the opportunities for the processing industry in integrating the human factors (HF) in hazard identification and risk assessment and suggest a practitioner's experiential approach to successfully manage the integration. It will, in detail, discuss at a conceptual and practical level the types of risk assessment techniques and tools, considering the often lack of effective evaluation of HF interaction with the dynamic process risks.

Any processing activity starts with the consumption of resources or raw material as feed which is then processed to deliver a desired final product. During the processing, energy, and utilities are consumed in one form or the other, residue (waste) is left for further processing/recycling/disposal, and in most of the cases gaseous emissions are left into the environment.

In most of the modern processing industries the materials being handled are hazardous due to their toxicity, flammability, corrosivity, radioactivity, or other hazardous chemical properties. Also, the processes themselves are becoming more and more complex and hazardous due to the nature of

Process Safety Management and Human Factors.
DOI: https://doi.org/10.1016/B978-0-12-818109-6.00006-X

61

reactions requiring high pressure and/or high temperature, vacuum, or cryogenic conditions. Hence, over the last few decades the processing industry has been focusing more on the process safety management (PSM) by developing management systems and deploying various hazard identification and risk assessment tools.

Major chemical disasters such as Bhopal gas tragedy and Piper Alpha incident triggered the early set of stringent industrial safety regulations by various countries such as the USA and the UK during the mid to late eighties of the last century. The early regulations requiring that industry must adopt a systematic risk assessment was first demanded by the government of the United Kingdom following the Piper Alpha disaster in 1987.

Major industrial accidents with catastrophic conequences over the last few decades forced governments to introduce suitable regulations that underscore the need for companies to identify potential human errors and to reduce the frequency and consequences of such errors as part of an overall PSM program. However, for those who are responsible for conducting or coordinating the risk assessment, the real challenge appears to be the level of awareness and understanding of the Human Factors, quite often 'the elephant in the room'.

This chapter describes an approach for integrating the Human Factors into the process hazard evaluation throughout its life cycle, that is, process designs, deviations/changes during construction and commissioning, and health, safety and environment (HSE) management system for safe operations during normal and abnormal situations. The risk management across the entire life cycle of any process facility should ideally include appropriate hazard identification and risk assessment requirements at each phase of the life cycle as illustrated in Fig. 6.1.

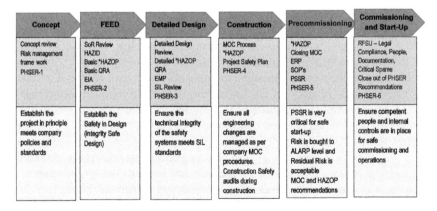

FIGURE 6.1   Typical HSE Risk Assessment/Studies in any major project life cycle.

Traditional approaches to the hazard identification and risk assessment revolve around deployment of standard tools such as hazard identification (HAZID) and hazard and operability studies (HAZOP), quantitative risk assessment, safety integrity level analysis, and layers of protection analysis (LOPA), etc. The human-machine interface (HMI), human-system interface (HSI), and HF in general are neither addressed during the design stage nor factored in the safe operating procedures/ standards within HSE management system.

Human factors as defined by the UK Health and safety Executive' in their HSG-48 document refer to "environmental, organizational and job factors, and human and individual characteristics, which influence behavior at work in a way which can affect health and safety". From process safety perspective, Human Factors are those aspects of the process and related systems that make it more likely for the human to make a mistake that in turn causes or could cause a deviation in the process or could in some indirect way lead to the increased probability of an accidental loss.

Addressing Human Factors during risk reviews especially in HAZOP of the design drawings as well as HAZOP of operating procedures can help identify gaps related to various Human factors which in fact can act as effective safeguards once they are properly addressed. This is fundamental for predicting where and why humans might make mistakes and for determining (qualitatively, at first) if the protection layers are adequate and effective against potential health and safety consequences caused by such errors should they occur. Unfortunately, many organizations do not fully analyze for errors during all modes of operation, and so many (in some cases, most) of the accidents that start with a human error are not predicted, as they should be. (For further reading on the subject, please refer Chapter 9.1 of CCPS/AIChE, Guidelines for Hazard Evaluation Procedures, 3rd Edition, 2008; and paper by Bridges, LPS/AIChE, April 2009, Optimizing Hazard Evaluations).

6.2 Human factors engineering in design and operational phases

Poorly designed Human-System Interface (HSI), for example, may make it difficult for personnel to detect changes in important parameters or to interpret the information displayed correctly. Difficult-to-use HSI may also frustrate the operators and inadvertently communicate a management message that accurate, timely human performance is not so important.

- HSI is defined as the technology through which personnel interact with plant systems to perform their functions and tasks. The major types of HSIs include alarms, information systems, and control systems. Each type of HSI is made up of hardware and software components that provide information displays, which are the means for user-system interaction, and controls for executing these interactions. Personnel use of HSIs is influenced directly by:
  1. The organization of HSIs into workstations (e.g., consoles and panels);
  2. The arrangement of workstations and supporting equipment into facilities, such as a main control room, remote shutdown station, local control station, technical support center, and emergency operations facility; and
  3. The environmental conditions in which the HSIs are used, including temperature, humidity, ventilation, illumination, and noise. There are three important goals to be achieved in the design and implementation of an HSI. These are:
     - **Design for operability** refers to designing the HSI to be consistent with the abilities and limitations of the personnel who will be operating it. Weaknesses in the design processes can result in an HSI that is not well suited to the tasks that personnel must perform to ensure plant safety, resulting in increased workload, decreased performance by personnel, and an increased likelihood of errors. For example, in one of the fatal incidents in an upstream oil company, a relief valve on a low-pressure separator actuated during apparently normal operations as a Pressure transmitter on that vessel failed closing the normal discharge valve and leading to a false signal to the control room. Operators verified that the separator pressure was normal, and in their haste to stop the release, they blocked-in the "bad" relief valve before unblocking the parallel relief valve. The separator immediately ruptured resulting in the death of two operators. In this case the HSI was poor as the design of the system and its configuration in the DCS and Control Room panels was faulty due to lack of interlocks on relief valve actuators.
     - **Design for maintainability** refers to designing the HSI and associated plant equipment to ensure that personnel can perform necessary maintenance activities efficiently. Weaknesses in the design process can result in systems that impose excessive demands on personnel for maintenance and, therefore, are prone to maintenance errors or problems with reliability and availability.
     - **Design for flexibility** refers to the way that changes, such as upgrades to the HSI, are planned and put into use. A new HSI component may require the user to perform functions and tasks

in new ways. Skills that the user developed for managing workload when using the former design, such as ways for scanning information or executing control actions, may no longer be compatible with the new design. The new HSIs must also be compatible with the remaining HSIs so that operators can use them together with limited possibilities for human error. Also, HSI modifications may not be installed or put into service all at one time, causing the user to adapt to temporary configurations that are different from both the original and final configurations. Weaknesses in HSI implementation can increase operator workload and the likelihood of errors.

- **Improving HSI:** Verify whether a HF engineering evaluation of the processes and control panels/rooms was performed; check whether all critical components are clearly labeled; check for ease of selecting the right device or component, whether the system feedback is clear and timely from the loops; and check if workers need to bypass interlocks often and whether there is adequate labeling of emergency escape routes, alarm panels, and safety critical equipment including color coding, tags, numbers, etc.
- **Check whether the equipment is poorly designed for human use**, for example, valve flanges or pipe joints inaccessible for unbolting or pressure gauges or flow meter at heights without any safe access, thus making it difficult for the operator to check them regularly during operations.
- **Good human factor engineering** can be as simple as "grouping" control elements by function: Management should ensure they have a minimum specification for HF engineering, from control panel layout to control room layout.

6.3 Task design

A task that is designed with the human limits in mind is much more likely to work effectively than one that assumes humans can and will "always" do what is written. The task must consider that humans think and remember and factor in prior data along with their previous experiences.

Complexity of task (procedure-based or a call for action): If the task is too complex, then humans can forget their place in the task, fail to understand the goal of each step or substep, or fail to notice when something is not going right. Complexity is difficult to predict (since it is not known when a human intervention will be needed), but higher complexity can increase error rates by 2–10 times.

Error detection and error recovery: Is there enough feedback in the process to allow the worker to realize (in time) that they made a mistake?

Have they been trained on how to reason through how to recover from mistakes they or others make? (Sometimes, doing a step too late is far worse than NOT doing the step at all.) Is there enough time available for the type of intervention necessary or is the environment where the task is to be performed (too noisy, too cold, to hot, too distracting)? And so on.

6.4 Procedures

Accurate, accessible, and usable procedures also play an important role in directing attention, and lack of accurate and easily accessible procedures can frustrate the worker and degrade their motivation. While risk reviews cover the detailed design parameters and piping & instrumentation diagrams (P&ID), no such review is typically conducted on the operating procedures and work practices.

It is suggested that simple techniques such as what-if or HAZID may help in integrating HF in noncritical operating procedures, while a HAZOP study may be quite helpful to address the HF integration for process safety related documentation, including safe operating procedures.

6.5 Human resources

Key guide words to be considered while carrying out HAZOP of P&IDs and SOPs under Human Resources are adequacy of manning levels, competence of the employees/contractors, level and quality of supervision, leadership commitment and organizational work culture. Performance management, worker engagement channels, and communication modes and their effectiveness are other critical factors to be considered during risk reviews.

Weaknesses in the personnel job performance evaluation and reward systems also may fail to communicate management expectations or may reward behavior that does not meet those expectations. If disciplinary actions are not perceived as being administered fairly, employee motivation to work productively and safely will be reduced.

Supervision: Supervision communicates and reinforces management expectations and establishes goals and requirements for task performance. Supervisory oversight may increase motivation to perform in accordance with expectations as well as detect and correct any errors that occur. Weaknesses in supervision, for example, may cause staff to choose production over safety goals in their work or to tolerate workarounds that may lead to errors.

Communication between workers: verbal and signal communication: Communication is the exchange of information while preparing for or performing work. Verbal communication occurs face-to-face, by telephone, sound-powered phones or walkie-talkies, as well as over public address systems. Written communication occurs, for example, through policies, standards, work packages, training materials, and e-mail. Communication involves two sets of behaviors:

1. creating and sending messages and
2. receiving and interpreting them.

Communication always involves at least two individuals, the sender and the receiver, and occurs:

- between individuals or within and among work groups;
- in meetings;
- in prejob or preevolution briefings; and
- during shift-turnover, successful communication requires several steps and involves many personnel including external agencies and contractors. If the communication is successful, the receiver interprets the message consistently with the sender's intended meaning.

6.6 Physical exposures

Exposure to extreme environmental conditions at the work place is one of the Human factors that could seriously impact the performance of individuals and more importantly their health and safety. The following environmental and physical conditions which could influence the Human factors adversely should be addressed at each and every phase of the risk assessments through out the project life cycle.

- Hand Arm Vibration: The effects of vibration depend upon its frequency and acceleration. Frequency is the number of oscillations (cycles) that occur in one second. Acceleration is the force, or intensity, of the vibration.
- Noise: Abnormal noise can cause errors, may disrupt communications, affect the ability to perform tasks, and annoy personnel.
- Exposure to heat or cold conditions: A common problem in many areas of a plant, such as the furnace or boiler rooms or steam turbine plants when the plant is operating. The extent to which workers will be affected by heat depends on many factors such as airflow, humidity, clothing, and level of physical activity and personal factors such as age, weight, acclimation to heat, physical fitness, and dehydration.
- Adequate lighting is required for accurate performance of nearly every task in a unit operation.

The organization must have engineering controls to help control each factor, but sometimes there may be no other choice but to rely upon administrative controls.

6.7 Fitness for duty

Fitness-for-duty issues include reduction in an individual's mental or physical capabilities due to substance abuse, fatigue, illness, or stress, which increases the likelihood of errors. Depending on the criticality of the tasks being assigned to an operator, suitable physical fitness requirements shall be identified and prescribed as mandatory requirement. The HAZOP study at both Design review and SOP review stages should include a review of such requirements and whether the fitness requirements are adequate to address the risk exposure. Types of possible impairments are:

- Physical attributes: strength, reach, eye-sight and color acuity, and hearing.
- Mental attributes: drug and alcohol (abuse); mental stress (on and off the job); fatigue: issues from on the job and off the job (especially control of hours per work-day and per work week).

Company programs that may be implicated in errors caused by personnel impairment include:

- Fitness-for-duty program: Company fitness-for-duty programs are primarily responsible for detecting and preventing impaired personnel from performing tasks that may affect public health and safety. Medical evaluations of personnel, behavioral observation programs, employee assistance programs, and drug and alcohol testing are used to detect impairment. Weaknesses in this program may allow impaired personnel to have access to vital areas in a plant where they could commit errors. One excellent starting point for a Fitness-for-duty subelement is the guidance provided in US NRC's 10 CFR 26 (2005).

- Overtime Policies and Practices: Most companies establish limits for work hours to reduce on-the-job fatigue. It has been shown that 17 hours of work without a break is the same as being legally drunk and at 10 days straight of 12 hours' work-days, the error rates for nonroutine tasks such as startup of a continuous unit can increase to 1 mistake in 5—10 steps (as opposed to the target of 1 mistake in 100 steps). Routine authorization for work hours more than those recommended may result in fatigued workers.

- Further, a practice of excluding training or meetings that occur outside of an individual's normal work schedule from work-hour limitations will also contribute to fatigue. In US NRC's 10 CFR 26 (2005), the guidance given for control of overtime hours is no more 72 hours of work per 6-day period, no more than 16 hours of work in 1 day within that period, and a minimum of 24 hours contiguous hours away from work within a 7-day period. The US DOT has even more stringent rules on limiting work hours and establishing required hours for recovery from fatigue.
- Shift scheduling: Shift scheduling may also affect the likelihood that personnel will show performance decrements due to fatigue. A change in the assigned shift or a rotating shift schedule will disrupt circadian rhythms and may increase the likelihood of errors.

6.8 Incident investigation

As the famous saying goes, "Those who do not learn lessons from the past are condemned to repeat them in the future." Although most companies may have some sort of an incident investigation system, but the associated tools may not be based on a root cause analysis (RCA) thus they may completely miss out the root causes! Even if the company has an RCA tool, they may lack the awareness of where HF fit into an accident sequence. This leads the investigators to often blame the "human error" as a root cause and stop at that casual factor without drilling down further to identify the root cause(s).

Open reporting of near-miss events, incidents and or other deviations from normal process conditions is essential for improving our understanding of Human Factors at work and effectively integrate them in our operating systems and management culture for sustaining an interdependent safety culture. However many organizations do not have robust reporting and monitoring system with emphasis on human factors perhaps due to lack of awareness and understanding of the criticality of Human Factors at work. Many companies also do not get nearly enough near misses reported (the ratio should be about 20–100 near misses reported per actual loss/accident; see CCPS/AIChE, Guidelines for Investigating Chemical Process Incidents, 2nd Edition, 2000; and paper by Bridges, 8th conference, ASSE-MEC, 2007, Getting Near Misses Reported).

For an effective integration of HF, proper conduct of HAZOP during the design review and HAZOP of Safe Operating procedures is essential and the lessons learned in the form of root causes from previous incidents should be captured during HAZOP review to prevent reoccurrence of such incidents in future.

6.9 Safety culture

The safety culture maturity plays a great role in process safety management. Hence it is imperative to consider the safety culture aspects while reviewing the hazard identification and risk review of project design basis and the relevant operating procedures.

Behavioral safety improvement programs are key enablers for nurturing an interdependent safety culture in any organization. These programs focus on identifying and correcting human behaviors that may result in adverse consequences through behavioral observation and feedback from supervisors and peers. Some programs also emphasize self-checking, such as DuPont's STOP program, FMC's START program, the Institute for Nuclear Power Operations' STAR program (stop-think-act-review), and PII's STAR Program (Safety Task Action Reporting), etc.

The level of effectiveness of such programs is another valuable indicator that can be considered while integrating HF into hazard identification studies and risk review activities.

6.10 Conclusions

For organizations that plan for an effective integration of Human Factors within their process safety management system, the author suggests them to consider the following approach which is developed based on an extensive experiential learning.

1. Develop a list of HF guidewords along with the potential deviations and their impact on the process/asset/people/environment.
2. Develop simple but clear procedures on how to integrate HF guidewords in the company's hazard identification and risk assessment procedures, for example, HAZID and or HAZOP studies. Some of the key HF Guide words which are explained already in this chapter are listed for ready reference:
   - HSI design: Complexity level
   - Staffing levels and competence issues
   - Task design and procedural requirements
   - Fitness for duty
   - Effective shift communications
   - Maturity of organizational culture
   - Learning from incidents
   - Engagement and empowerment levels, etc.
3. Follow the Risk-Based Process Safety standard from AIChE/CCPS, and make sure the elements that control HF are fully implemented

(including all advice on HF given in this paper), especially for the elements of (a) process safety culture, (b) Workforce involvement, (c) training and performance, (d) operational readiness, and (e) conduct of operations.

4. In addition, companies may supplement this approach by implementing a PHA/HAZOP management system that defines how PHA teams should address HF and how teams should analyze specifically for human errors of not following operating procedures. Companies should also ensure that incident investigation teams are trained on HF to help them correctly identify the HF weaknesses that may led to human errors or component failures.

7

Inherent safety impact in complying process safety regulations and reducing human error

Risza Rusli and Mardhati Zainal Abidin

Centre of Advanced Process Safety, Universiti Teknologi Petronas, Seri Iskandar, Perak, Malaysia

7.1 Introduction

The conventional practice of process safety management consists of the tremendous interdependency of the various elements such as identification of hazards, evaluation of risks and implementation of controls are heavily influenced by human factors. To ensure the success of process safety management for complex and dynamic environments where there are many variables and factors that can go wrong, a proper training program is needed for industry. While the training involving hard skills, such as equipment operation and understanding of regulations, can increase the safety level of a plant, major accidents have been recurring over the years (e.g., Flixborough accident, Bhopal accident, and Texas City accident). Other than human nature itself, a lack of focus on the key soft skills such as communication and crisis management that allow operators to respond effectively during incidents contributed to these accidents. The best way to resolve this issue is by proactively removing the source of the hazard itself through inherent safety implementation. This chapter will address the statistics on human error that lead to the accidents in chemical process safety, the basic inherent safety principles in process safety management, and how the implementation of inherent safety can reduce human error during operation. The benefits of inherent safety applications toward compliance of regulations, including evaluation of other type of control measures and training required to manage hazardous process will be demonstrated and discussed using case study.

7.2 The causes of accidents in chemical process industries

Chemical process operation are intensive activities requiring high man−machine interactions in challenging complex sociotechnical system due to the evolving conditions such as workplace process parameter, noise and vibration, and workload and stress. For example, extreme working condition affects operators' performance, increasing the chances of error, and, consequently, can cause injuries or fatalities to personnel. In the 284 cases studied by Kidam and Hurme [1] on the six most commonly process equipment involved in accidents, 623 causes to the accidents were found. Fifteen different types of accident contributing factors were discovered. As seen from Fig. 7.1, overall, human and organizational causes is the largest category (20%). Accident causes are often classified as technical, human, and organizational causes. The division between these is not clear, since many technical causes involve human related aspects such as design, installation and service errors or faults in the operator−technical interface. The latter is not an error in a technical sense but causes operators to make errors in operation. It is also typical that both technical and human and organizational causes contribute to accidents at the same time.

7.3 Inherently safer design in process safety management

Inherently safer design (ISD) was first introduced by Professor Trevor Kletz around four decades ago as a proactive approach in avoiding process safety risks. The main principle of ISD is to eliminate hazard from the source rather than accepting and trying to control the hazard. This approach can be materialized based on four main strategies: minimize, substitute, moderate, and simplify [2]. The principles emphasized on the reduction of hazard by reducing the quantities, replacing with less hazardous materials and operating the process in the less hazardous conditions. ISD concepts will enhance overall risk management programs, by reducing frequency or consequences of potential accidents through the four principles [3]. Exploring ISD may require more resources during the early stages of development. However, the results

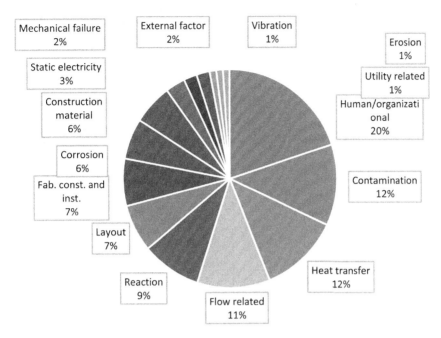

FIGURE 7.1 Contributors in equipment-related accidents. *Data from: K. Kidam, M. Hurme, Analysis of equipment failures as contributors to chemical process accidents, Process Saf. Environ. Prot. 91 (1−2) (2013) 61−78.*

minimize or eliminate safety mitigation devices and the costs of maintaining them, as well as reduce the possibility of incidents. Inherently safer considerations may reduce the life cycle cost of the process as shown in Table 7.1.

7.4 Reduction of human error through inherent safety

Most of the previous works have highlighted the benefits of ISD in terms of reducing the risk and requirement of safety system, that is, controller and safety measures. Nevertheless, the benefits of ISD implementation in a broader and more detailed perspective have also been noted. According to Gupta et al. (2003) [5], the implementation of ISD allows the cost of a new plant that handles hazardous operations to be reduced up to one-third to one-half, or even more. This is because ISD focuses not only on the reduction of a safety system but also on other benefits, such as the reduction of risk assessment requirement, the reduction of emergency response plan requirement, the reduction of maintenance requirement, the reduction of training requirement, and

TABLE 7.1   Usual plant design activity and ISD review criteria in chemical process development and design [4].

Design phase	Process design task	ISD review criteria
Research and development (R&D)	Reaction concept selection • reaction chemistry • reaction kinetics • synthesis routes • process conditions Thermal analysis • calorimetric test • runaway reaction Examination of materials and chemical hazards • fire and explosion • toxic release • reactivity and stability • incompatibility Process scale-up bench and pilot scale test	• ISD for reaction systems that are focused on the inherent hazards of chemicals and their reactions potential at selected process conditions. • Proper safety investigation and laboratory test of process materials used, including feedstock, intermediate impurities, by products, and services medium. • Identification of the side effects of normal process condition and possible deviations to the material of construction. • Detailed study on the potential of unwanted reactions and the ways to control the reactive conditions, such as by using special additives and inhibitors, or by using solvent to quench or stop the reaction. • Examination of the potential internal/ external process contaminants and their impact to the reaction system. • Detailed analysis on the effect of temperature, pressure, humidity, long storage (shelf- life), and process deviation on the chemical properties of process materials. • Detailed study on the heat generation, thermal safety and potential of runaway at normal and abnormal process conditions. • Proper testing and modeling of a scaled-up reactor system.
Preliminary engineering	Process concept selection • mode of operations • process flowsheet • material balance • energy balance Unit operation selection • types of equipment • process condition • sizing Process modeling and simulation Logistics and material flow analysis	• ISD for main process and sub- process that focus on the inherent hazards of chemicals, their interactions at selected process conditions, and types of unit operations. • Designing of a smaller chemical process plant to reduce the inventory of hazardous materials. • Proper test and modeling of process equipment for safe design and scale-up. • Selection of appropriate types of equipment that are simple and easy to operate for the reduction of complexity and human error. • Constant use of and timely updating the design standard and checking of the local legal requirements.

(Continued)

TABLE 7.1 (Continued)

Design phase	Process design task	ISD review criteria
	• storage condition • transfer and handling	• Identification of the potential operational problems based on the characteristics of process material by choosing an experienced design team or by appointing reliable, external design specialists.
Basic engineering	Process equipment design safe operating limits • protection material construction • chemical and mechanical spec • equipment layout and configuration • feeding system Process system engineering automation and instrumentation • protection and mitigation Plant-wide layout physical arrangement • equipment siting and material flows Utilities set-up • heating/cooling system heat transfer medium selection	• ISD for process and unit operation design that are focused on safer, simpler, and robust design specifications. • Improvement of the integrity of the process equipment by using chemically-resistant and mechanically-robust construction materials. • Designing of process equipment based on worst-case scenario and employment of suitable protection system wherever appropriate. Analysis of chemical incompatibility and possible operational problems of the utility system by reviewing their chemical and physical properties and process conditions. • Selection of the correct supply energy for the specific application. Foreseeing of potential internal and external process contamination through process connectivity and equipment/pipe sharing. • Selection of robust, reliable, simple, and user-friendly control system. • Use of simpler control philosophy particularly in emergency situations. • Designing of a simpler chemical process plant to limit the process interaction during process deviation and upset. • Provision of adequate separation between process units, storage area, and other buildings. • Designing of a simple and robust piping system by eliminating unwanted components, dead-end, bypass, and standby pipelines.
Detailed engineering	Detailed process description on: • normal operations • process start up and shutdown	• ISD for detailed process equipment designs that is resistant to process deviations, emergency, and accidents. • Detailed study and modeling of flow-related problems using a simulator and advanced metallurgy data.

(Continued)

TABLE 7.1   (Continued)

Design phase	Process design task	ISD review criteria
	• automation and control • preventive maintenance and services • emergency preparedness • safety manual • code of practice • safe system at work • management system	• Special technical and operational consideration for layout and mechanical performance under severe process conditions and unwanted events (i.e., internal/external fire or explosion). • Selection of simpler, robust, and user-friendly "buy item." Reviewing of the conductivity of the material of "buy item" to limit the generation and discharge of static electricity. • Selection of a safer material and working method for cleaning, servicing, maintenance, and repair work.

the increase of community acceptance. These elements are elaborated as per the current issues as follows:

- Regulatory compliance: Compared to the inherently safer process, operating a hazardous process is subject to stringent requirements to ensure a safe operation. Failure to comply with the regulatory requirements is reciprocated with a hefty fine. Nevertheless, cases related to the violation of regulations are common in the industry. In 2009, the US Department of Labor's Occupational Safety and Health Administration (OSHA) issued more than $87 million penalties to BP Texas due to the violation of process safety management. The amount was noted to be the largest recorded in OSHA's history.
- Risk assessment: The Malaysian regulation requires a detailed quantitative risk analysis of hazardous process for the submission of a safety report to the regulatory body. To do so, the personnel need to be trained to perform risk analysis, and the plant owner must consult a competent person. However, the service of a competent person does not guarantee the acceptance of the report; it could, nevertheless, reduce the number of reports that are rejected, require modification, or require correction. On-site and off-site emergency response plans need to be prepared, which require the involvement of civic authorities. The plans need to be updated regularly and the designated person needs to be trained.
- Safety system: The equipment including active (e.g., leak detectors and sensors, alarms and control equipment) and passive (e.g., dikes, catch pots) need to be installed in the plant to limit the consequences of major accident on human health and environment. This requirement also covers the installation required to handle off-spec

materials and spills. To ensure that the monitoring process and controllers perform as intended, the software must be reliable and handled by trained personnel.

- Emergency preparedness: For hazardous site, the emergency response plan needs to be prepared and updated regularly. Trainings and mock drills are required and need to be conducted regularly. Other than that, on-site medical facilities are required and personal protective equipment (PPE) must be provided to the workers. The PPE and medical facilities need to be replaced within time intervals. Under emergency response planning, the internal reporting systems for all near- miss incidents and accidents are required and must be handled by trained personnel. A safety-and-health committee needs to be established in the company, and the meeting among the committee members must be held in regular intervals.
- Plant layout and arrangement: Any expansion/modification performed in the hazardous plant must be approved by authorities prior to the modification. The arrangement of buildings, layout, design, and construction must follow a standard code. Usually, a detailed plant layout that shows the arrangement of equipment must be submitted to regulatory body for approval prior to the installation or modification.
- Maintenance: All equipment, safety system, and instruments of the hazardous plant must function as they should, hence requiring regular maintenance. Other than the cost required to train maintenance personnel/hire maintenance personnel, the cost for maintenance itself is substantial. Details of the maintenance need to be recorded and an inspector from the regulatory body has an authority to access the records whenever required.
- Transportation, packaging and labeling: The packaging, labeling, and transportation of hazardous material need to comply with the strict procedure and guidelines. The driver must be trained and be capable of handling hazardous materials, and the off-site emergency plan needs to be prepared in the case of material release that is caused by accidents during transportation.
- Relationship with civic authorities: Because the process of managing hazardous process requires the commitment and involvement from civic authorities or inspector from regulatory bodies, good relations with them must be maintained. The staff must (1) be prepared to host their visit (2) prepare the equipment so that it is ready for inspection and (3) manage the follow-up action. In Malaysia's regulation, there is a clause that allows the inspector to cancel the inspection if the process/equipment is not prepared for inspection. In such occurrence, the company needs to arrange for another inspection date. The cost for additional inspections must be borne by the company.

- Community acceptance: Good relationships with media and community need to be maintained. In recent years, many hazardous operations have been canceled due to the poor relationship between the plant owner and the media and community. The community has the right to (1) know the nature of the major accident hazards, including their potential effects on human health and the environment, and (2) the details of the main types of major accident scenarios and the control measures to address them. With the increasing awareness related to safety-and-health environment, operating hazardous operation requires extra efforts and funding to educate the community.
- Cost: Operation using pure hazardous raw material might incur high cost. The building that accommodates hazardous operations need to be equipped with fireproofing, fire escapes, and sprinkler systems, to list a few. Higher separation distance is required between equipment as per the standards, thus additional cost for piping and land are needed. The equipment to handle hazardous material and transportation need to be fabricated based on higher standards and thus will increase the cost.

By implementing ISD, the needs of the elements to manage a hazardous process as discussed above can be eliminated/reduced to the minimum level as the process becomes safer. Improvement can be achieved because the operating cost and maintenance cost can be reduced. Other than that, a safe process operation can be accomplished with the reduction of human error since unnecessary procedures for managing a hazardous process can be reduced/eliminated. With the implementation of ISD, a safe environment to the society and environment sustainability can be guaranteed, and subsequently increase the social status of a company.

7.5 Case study

In this section, the identification of ISD alternatives for ammonia storage using the Inherent Safety Benefits Index (ISBI) method outlined in [6,7] will be demonstrated and the impact toward the reduction of human error will be discussed. The ISBI which is proposed to identify the best ISD alternative (1) with the lowest severity of accident using Damage Index (DI) and (2) that requires minimum efforts to manage a hazard using Hazard Management Index (HMI). The HMI was developed based on three regulations; the Control of Major Accident Hazards Involving Dangerous Substances (COMAH 2015), the United Nations Recommendations on the Transport of Dangerous Goods (UNRTDG), and the Globally Harmonized System of Classification and Labeling of

Chemicals (GHS), to show the benefits of ISD in terms of the reduction of regulatory requirement to manage hazardous process.

7.5.1 Ammonia

Ammonia is a toxic material that exists in a gaseous phase at ambient temperature. In the common practice, ammonia is liquefied either by cooling or compression for ease and economical transportation and storage. In the work of Roy et al. (2011) [8], pure ammonia was stored in a liquefied form in a high-pressure storage. Such a design is hazardous due to the high concentration of toxic material and high-pressure operation. The hazard identification of the system is shown in Table 7.2.

7.5.2 Inherently safer design alternatives for ammonia storage

Three ISD alternatives were proposed according to the common industrial practice as performed by various studies [9–11]. The number

TABLE 7.2 Hazard identification results for the base case [7].

Hazards identification	Description
Material	• Anhydrous liquid ammonia
Undesirable consequences	• High toxicity level • Exposure of vapor or liquid has the potential to cause serious injury or fatality and environmental contamination
Material properties, system, process, and plant characteristics	*Material properties* • Color: colorless • Standard: gas • Relative density, gas-0.6 (air = 1); relative density, liquid-0.7 (water = 1) • Vapor pressure: 124 psi at 20°C (68°F) • Boiling point −33°C • Solubility in water, completely soluble • Percent volatility: 100% • Lower explosive limit (LFL) −15%; upper explosive limit (UFL) −30% • Immediately dangerous to life and health (IDLH)-300 ppm • Acute exposure guideline levels (AEGL) for 10-min exposure durations; AEGL-1−30 ppm AEGL-2−270 ppm AEGL-3−2700 ppm *System/Process* • High-pressure storage

and capacity of tanks for the ISD alternatives were estimated based on the limitations addressed in previous studies by [10,12,13]. Details of the ISD alternatives are given in Table 7.3. Alternative 1 offers an opportunity for hazard reduction based on the moderation principles, particularly by operating the hazardous system with less hazardous conditions. The high-pressure tank was replaced with a cryogenic storage system in which the liquefied ammonia was stored using a refrigerant at ambient pressure. Alternative 2 also offers hazard reduction via the moderation principles. It offers an opportunity for hazard reduction through a dilution approach in which aqueous ammonia with 29% concentration was stored in a high-pressure vessel. Alternative 3 also offers hazard reduction potential using the moderation principles, particularly by replacing the high-pressure tank with an atmospheric storage system, which stores a lower concentration of ammonia (19%) at ambient temperature. The results obtained for the four alternatives are summarized in Table 7.4.

The impact of ISD modification on the potential of accident DI was influenced by inventory, phase, pressure and vapor density. In the case of both inventory and pressure, the reduction of these values will reduce the damage radii, one by reducing the quantity of material release, the other by reducing the rate of release. Temperature and material dilution played an important role in reducing the vapor density. The release of ammonia from a high-pressure storage (base case) led to the depressurization process, which resulted in a two-phase flow in which part of the fraction formed a pool and the other part formed an aerosol. The aerosol then increased the cloud density and behaved as heavy gas and slumped to the ground. The heavy gases then caused a build-up of lethal toxic load closer to the ground for a longer duration

TABLE 7.3   ISD alternatives for ammonia storage [6,7].

Option	Base case	Alternative 1	Alternative 2	Alternative 3
Types of Storage	High-pressure Storage	Atmospheric Storage	High-pressure storage	Atmospheric Storage
Substance	Anhydrous ammonia	Anhydrous ammonia	Aqueous ammonia (29%)	Aqueous ammonia (19%)
Pressure (kPa)	1274.86	101.33	108.27	101.33
Temperature (C)	25.00	−33.40	25.00	25.00
Mass (kg)	40,000.00	160,000.00	95,663.00	881,118.00
No of unit	4	1	6	1

TABLE 7.4   ISBI ranking for ammonia storage tank [6].

Option	Base case	Alternative 1	Alternative 2	Alternative 3
DI	68.01	9.37	35.78	11.80
HMI	2	2	1	1
ΔDI	0	6.26	0.90	4.76
ΔHMI	0	0	1	1
ISBI	0	0	0.90	4.76
Rank	4	3	2	1
	Base case	Less significant positive impact of ISD Implementation. Reduction of either one - accident severity or regulatory requirements	Significant positive impact of ISD Implementation. Reduction of both accident severity and regulatory requirements	Significant positive impact of ISD Implementation. Reduction of both accident severity and regulatory Requirements

than do lighter gas. Combined with the gravitational impact, the heavy gas slumped to the ground, spread horizontally, and increased the hazard zone. By lowering the temperature, (1) the vapor pressure can be reduced; (2) and no initial flash of liquid can vaporize in case of a leak when a material is stored at or below its atmospheric pressure boiling point; and (3) a two-phase flashing jet can be eliminated. Material dilution, on the other hand, not only can lower the initial atmospheric concentration, but also gives a similar effect in reducing vapor pressure hence resulting in the reduction of release rate. Thus, the impact of ISD moderation principles via dilution and temperature reduction can significantly reduce the dispersion of ammonia gas.

Other than the potential for fatalities, ammonia release will cause long-term environmental pollution. Ammonia release in water will result in nutrient pollution, which in turn causes a eutrophication phenomenon in a natural water source. This phenomenon will result in excessive plant growth or the decay of certain species as well as the reduction of water quality. Excessive plant growth will reduce the content of oxygen in water and affect the aquatic life. Release of ammonia in the soil will change the pH of the soil to an acidic value. The soil needs to be treated using a leaching process for plantation. Fertilization of vegetation can occur and will cause the spur of weedy growth while choking out native flora and wild flower. These issues will result in imbalanced ecological system. More than that, the combination of ammonia with nitrogen oxides (NO_x) with sulfur oxides (SO_x) emissions will form fine particulates that will cause haze. The larger area affected by the release of ammonia will result in higher DI for the environment. These results attest that the potential of severity can be reduced using an ISD approach, which are the moderate and simplify principles for this case. The positive impacts of these two principles appear to override the negative impact brought by additional inventory.

The HMI outcomes from this approach indicates that only base case and Alternative 1 that stored anhydrous ammonia were included in the enforcement of lower-tier COMAH regulation. Alternative 2 and Alternative 3 were excluded from COMAH enforcement because aqueous ammonia is not listed under the named substance category. As indicated from the results in Table 4.21, Alternative 1 (single unit with 160,000 kg capacity) and base case (four units with 40,000 kg capacity each) obtained the same HMI in accordance to the COMAH where the total amount of hazardous substance exists in the establishment must be considered while assigning the index.

In terms of regulatory requirements, Alternative 2 and Alternative 3 appear to be more favorable because they are exempted from the fulfillment of COMAH, whereas Alternative 1 and base case need to fulfill the lower-tier COMAH regulations. According to the COMAH guideline, if Alternative 1 or base case were to be selected, then the following notifications must be prepared and submitted to the authorities: detail information on the process, types, and quantities of hazardous substances under possession, the person in charge, and environmental and other factors that are likely to cause a major accident. Other than that, the Major Accident Prevention Policy (MAPP) must be prepared and updated, within a time interval, by the participation of authorities and other establishments; and all the requirements for packaging, labeling, and transportation must be fulfilled. In conclusion, operating a hazardous operation requires extensive efforts in managing the hazard.

In addition to the resources that must be spent to fulfill all the aforementioned requirements, additional requirements often lead to errors because the conditions might lead to repetition of the tedious process or worst, become the cause of accident. In the case a process is classified as a major hazard establishment with HMI value = 3, the requirements outlined in COMAH need to be fulfilled. One of the requirements is to prepare a safety report.

[13] S.D. Chavan, Cryogenic storage of ammonia, Chem. Ind. Digest. (2012) 76−80.

[14] T. Britton, Lessons learned about preparing comah safety reports, in Symposium Series No 149, 2003, pp. 55−67.

[15] S. Mustapha, I.M. Zain, Safety report:maintaining standard and reviewing, in Symposium Series No 149, 2003, pp. 87−101.

[16] A. Ramzy, After protests, chinese city halts chemical plant expansion, TIME, 2012. [Online]. Available: <http://world.time.com/2012/10/30/after-protests-chinese-city-halts-chemcial-plant-expansion/> (accessed 29.12.16.).

[17] Bayer Crop Science, The Use and Storage of Methyl Isocyanate (MIC) at Bayer CropScience, National Academies Press, 2012.

[18] G.C. Ta, M. Mokhtar, A. Mohd Mokhtar, A. Bin Ismail, M.F. Abu, Analysis of the comprehensibility of chemical hazard communication tools at the industrial workplace, Ind. Health 48 (2010) 835−844.

Asset and mechanical integrity management

Hirak Dutta

Formerly with Indian Oil Corporation Limited; Oil Industry Safety Directorate. (OSID); Ministry of Petroleum & NG; Nayara Energy Limited; Indian Society of HSE Porfessionals, Uttar Pradesh, India

8.1 Preamble

Major incidents continue to strike the oil and gas industry at regular intervals. A closer look and analysis of the incidents depict that the root cause of many of the incidents are alike and follow a pattern. Though the learnings from each of the major incidents are circulated, documented, and shared, failures of similar nature continue to happen.

Process safety failures are relatively rare, but their consequences are severe. The Bhopal Gas tragedy and the Piper Alpha disaster many decades ago to the more recent Deepwater Horizon, United States (2010), Buncefield (Hemel Hampstead) Terminal Fire, United Kingdom (2005), Jaipur Terminal Fire, India (2009), etc., are some of the sad reminders of the catastrophic process safety failures.

For the loss of property, loss of lives, and damage to the environment, process safety incidents can destroy the reputation of the organization. The customers, stakeholders, and the society at large get significantly impacted by these incidents. The biggest setback of any major process safety incident is that it demoralizes the workforce.

8.2 Process safety model

The hydrocarbon industry handles highly flammable petroleum products, toxic chemicals, and catalysts. Its processes are complex and

Process Safety Management and Human Factors.
DOI: https://doi.org/10.1016/B978-0-12-818109-6.00008-3

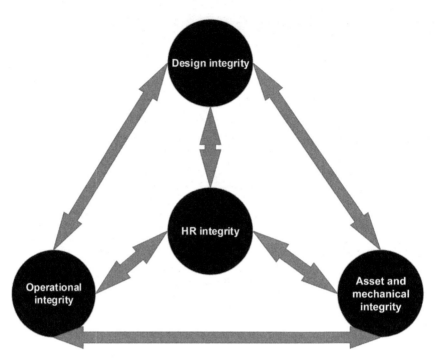

FIGURE 8.1   Process safety model: hydrocarbon sector.

operate at high temperature and pressure. The hydrotreating reactions in refineries which take place at high pressures and temperatures in the presence of hydrogen make the process even more hazardous. Thus, ensuring safe and reliable operation is a challenge before the industry.

The process safety model (PSM) that was developed many years ago is followed in the hydrocarbon industry today to ensure smooth, reliable, uninterrupted, and around-the-clock safe operation, and encompasses four pillars that include design, operations, maintenance, and asset integrity and human resources. See Fig. 8.1, which shows the PSM for the hydrocarbon sector.

8.2.1 Design integrity

Design integrity is the corner stone of PSM. Uninterrupted safe, smooth, and reliable operations cannot be achieved unless the design of the plant, be it a refinery or petrochemical unit or a POL terminal, is robust. It is therefore of paramount importance that process flow diagrams (PFD), general arrangement (GA) drawings, and piping and instrumentation diagrams (P&ID) are prepared and various layers of protection are incorporated at the nascent stage of design itself. This creates what we refer to as inherently safer design.

The integration of safety interlocks, distributed control systems (DCS) based instrumentation, process logic controls (PLC) based emergency shutdown systems, and other automations are necessary to ensure that design of the plant is fail-safe and the requirement for human interventions are minimized. Similarly, the equipment that are being procured must meet the relevant standards conforming to API, ASME, NFPA, OSHA, and other good industry practice specifications and regulations.

8.2.2 Operational integrity

The second pillar of PSM is operational integrity. One must have an in-depth knowledge of the "nitty-gritties" of plant operation. How to start up or shut down a plant? How to handle the emergency situations? How to manage and prevent runaway reactions/excursions in complex catalytic reaction processes?, and so forth. It is essential to prepare standard operating procedures (SOPs) which must be followed and shared with the personnel working at shop level to prevent any untoward incident. It must be ensured that SOPs are not violated.

As such the operational discipline in a plant must be instilled to ensure that the operating procedures which were also developed as part of the hazard and operability studies (HAZOP) are implemented.

8.2.3 Asset and mechanical integrity

The third pillar of PSM is asset and mechanical integrity. If the philosophy of an organization is based on breakdown maintenance—safety and reliability cannot be ensured. Reliable organizations have moved away from breakdown and reactive maintenance systems to preventive and predictive maintenance systems.

High reliability organizations (HROs) depend upon risk-based inspections and reliability-centric maintenance practices. The state-of-the-art upkeep practices ensure that every circuit and piping network in the plant are inspected and based on inspection observations, necessary mitigation measures are taken to ensure safe operation and a continued reliability of the plant.

Asset and mechanical integrity applies to all parts of the plant and its systems including pressure vessels, drums, columns, reactors, piping, rotating, and stationary equipment. As more and more automation has been introduced to plants, the reliability of the electrical/electronic and instrumentation systems and their integrity has become more and more significant as part of the holistic asset and plant reliability programs.

8.2.4 Human resources

Finally, people are at the very center stage of process safety management. It is people who drive the process and technology. Questions that management must ask themselves: Are they properly trained and educated? Does the organization believe in continuously updating their knowledge and skills, that is, is this a learning organization?

So, the focus is always on the learning curve and learning processes related to improving the overall competency of those who are responsible for running the plant. However, what kind of operating culture and as such we ask are the employees in the organization motivated? Is the climate conducive for fostering creativity and innovation? Are the suggestions of the ideas of employees executed in the company? And is there a fair and just safety culture which holds people fully accountable yet provides all the required knowledge and skills to perform safely all the time? These are some of the prerequisites for achieving process safety excellence within an organization.

8.3 People—process—technology alignment to achieve process safety excellence

To help prevent incidents and create a process safety excellence culture and organization that can continue to perform well over extended periods of time, an alignment between the processes, the technology and the people within an organization is key.

In saying this, we still find that accidents take place in continuous process industry.

Upon analysis of major global incidents indicate that accidents take place due to the following broad reasons:

- Focus on personal safety (e.g., zero-LTIR regimes) but not as much on process safety performance indicators.
- Lack of operational discipline in implementing the systems and procedures.
- Focusing more on the lagging indicators in the EHS KPIs rather than proactive leading indicators, which help assess the amount and quality of efforts in leadership preventing incidents.
- Inadequate risk control, for example, dysfunctional barriers.
- Ignoring the early warning signals.
- Faulty work practices/work ethics.
- Failing assets in terms of their integrity due to age and fatigue from operations.

In more general terms one could also say that the industry is not regrettably learning effectively from the lessons from previous incidents as the incidents keep on repeating themselves in different geographies but with similar root causes (Fig. 8.2).

We continue to remain preoccupied with and highlight how many accident-free man-hours of operation are achieved. Our preoccupation with the zero-LTIR regime and counting on lagging indicators hinders us from analyzing the leading indicators effectively. According to the research, organizations devote 80% of its time in reactive management such as maintaining and updating hazard register, accident reporting, etc., and only 20% of time is devoted to proactive issues like work methods, safe work practices, etc. As such the development of the API 754 was a critical development as a guidance for the industry to drive the culture of process safety indicators' best practice to balance between process safety and personal and occupational safety indicators [1].

Too much high concern on production and organizations being overtly occupied with achieving higher than design throughput is being penny-wise and pound-foolish. To achieve higher crude oil processing rate in primary crude oil distillation units, process heaters are subjected to hard firing. This results in skin temperature of the heater tubes overshooting the safe operating limit. Continuing operation with such extreme conditions could ultimately result in furnace tube failure and accidents thereof.

In India, for example, almost 85% of the incidents take place due to lack of operational discipline in terms of adherence to SOPs, violation of

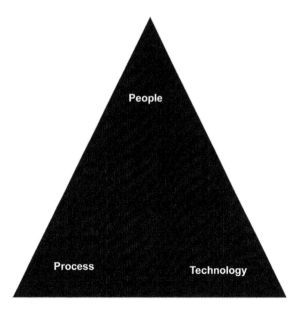

FIGURE 8.2   Alignment between people, process, and technology.

work permit systems, poor maintenance practices, and poor/inadequate supervision.

A sound asset integrity management program (AIM) can prevent unsafe situation(s) and thus avert accident(s).

8.4 Asset integrity management

One of the vital facets of ensuring safety in operations is asset integrity. An effective AIM program aims at establishing well-written accurate procedures that helps reducing human errors. AIM programs reduce the likelihood of failure by undertaking various quality assurance and preventive/predictive maintenance practices. Potential failures can also be identified and eliminated by carrying out time-bound/periodic inspections and tests and replacing the defective materials, safety devices, and so on.

Setting up additional facilities in the plant for operational flexibility or improving distillate yield or lowering energy intensity are desirable but the challenge is to maintain the healthiness of such facilities. This calls for, besides others, a proper inspection and maintenance regime for all the physical assets.

At locations, wherever the extant data on health of critical equipment or pipings are not available, a database on inspection must be prepared and completed expeditiously. Can an installation be called healthy unless proper inspection of drilling rigs, facilities at platform, critical equipment, pipe, and pipefittings are periodically carried out? Maintaining data on the current health of equipment is a must and assumes paramount importance.

The rotary equipment in any refinery, petrochemical plant must be categorized into semicritical, critical, and supercritical categories. The AIM plans are developed for critical and supercritical equipment. The signature analysis based on vibration readings of thrust bearings, journal bearings of high speed rotary equipment like recycle gas/makeup gas compressors, turbines are carried out and proper action is taken to prevent failures. HROs rely on online vibration monitoring probes to constantly monitor the health of the equipment and study through instituted logarithms of the performance of the equipment over time and benchmarked with optimized operational data benchmarks.

Failure of furnace tubes in a plant quite often leads to a major accident. To monitor the health of furnace tubes in high temperature furnaces besides measuring skin temperature and relying upon temperature indicators: an AIM program suggests measuring the skin temperatures of the furnace with infrared guns.

Similarly, to ensure good health of various other static equipment like columns, tanks, heat exchangers, other types of vessels, piping, etc., and the AIM program suggest a broad range of activities through visual inspection, thickness checking, detailed condition monitoring like acoustics emission tests, eddy current techniques, etc.

In more upstream oil and gas operations, the maintenance and certification of blowout preventers and other good control equipment must be carried out by original equipment manufacturers. The health of drilling rigs needs to be monitored and always maintained in a fit-for-purpose condition. The dynamic positioners with redundancy must be fitted to the service vessels. Proper healthiness of the service vehicles must be ensured, and obsolete ones must be phased out from service. Another important feature of maintaining good health of platforms is painting. The platforms operate in a saline (high chloride) environment and thus application of paint and its proper inspection is of significant relevance.

The practices of mean time between turnarounds (MTBT) have been improved in recent years to 4 years without taking into consideration that many undeclared shutdowns have taken place, and this must be shed away with. Delays in handing over of tanks to maintenance and inspection (M&I), or postponing annual turnaround to keep the production levels high must never be considered a prudent step. The oil companies must strictly adhere to the shutdown slates, put much emphasis on reliability improvement measures through reliability-centric management (RCM) and risk-based inspections (RBI) techniques to mitigate not only unsafe situations but also to improve productivity and profitability.

The challenge for effectively managing the health of old aging assets is even more. The design features, the safety instrumentation systems, the piping and piping network, the rotary and static equipment in the extant installations must be properly maintained through various available state-of-the-art measures to ensure that they are kept always in fit-for-use condition. The RBI-based approach focuses on such assets with high risk of failures through a systematic analysis on rate of corrosion and other damage mechanisms and suggests appropriate inspection techniques. This thus enables action to be taken to prevent failures.

In this chapter, next, we discuss a few case studies to highlight not only the failures and the lessons learned, but highlight the human factor failures in acting correctly and on time. The case studies highlight real cases and as we said in the outset of this chapter, we seem to not learn well from our previous incidents. We start off with a relatively recent case study of the failure of the Chevron-operated Richmond Refinery, just outside San Francisco, United States, and this is discussed later in the next section.

8.5 Industry case studies

8.5.1 Case study no. 1: Gas oil draw-off line failure at Richmond Refinery [2]

The fire in the gas oil (GO) draw-off line from the crude distillation unit at Richmond Refinery, California, happened besides others due to impatience to resume production postunit shutdown. The Inspection Department observed, during maintenance shutdown, considerable thinning of the GO draw-off line—necessitating line replacement. But the replacement of the line would delay the start-up and consequent production loss. It was decided to resume production.

Apropos start-up, GO leak was noted in the insulated GO draw-off line; GO is drawn off from the column at around 300°C. Maintenance crew wanted to have a close look at the exact location of the leak. While they were removing the aluminum-clad insulation with metallic hook, the fire took place due to friction. Smoke emanating from the incident surrounded the entire refinery and neighborhood. Luckily there was no fatality but 1500 residents in the neighborhood underwent treatment at hospital (eye irritation, etc., caused from the emanating smoke). The unit remained shut down for a considerable period. After the incident, it was found that there was 70% metal loss in the GO draw-off line which happened due to sulfonic acid corrosion (Fig. 8.3).

The unit was designed to process low sulfur-crude (S-crude) but subsequently changed over to processing high S-crude to improve gross refining margin (GRM). The original specification of the piping material for the GO draw-off line was carbon steel. High S-crude processing demanded changing the pipe material from carbon steel to higher metallurgy alloy steel. The leak in the GO draw-off line is attributed to a phenomenon—sulfonation corrosion, that is, S-corrosion—at high temperature in C-steel piping. It is worthwhile to mention that inspection reported considerable thinning of the C-steel pipeline.

The plant failure could have been avoided with the use of AIM program. RCM is a powerful tool that identifies the various potential sources of failures and its appropriate management thereof. Engineers drawn from multidisciplinary groups must work together to identify the systemic weaknesses and recommend measures to overcome the failure.

A multidisciplinary group of engineers drawn from chemical, metallurgical, mechanical, and instrumentation disciplines could have identified the weaknesses in the Richmond Refinery. The RCM team could identify the impact of processing low sulfur vis-à-vis high S-crude oil and the impact of corrosion in GO draw-off line. The sulfonation phenomenon viz. corrosion in the GO line at high temperature due to

FIGURE 8.3 Photo of the fire in gas oil draw-off line—hours after the incident.

presence of high sulfur could have been identified by the metallurgist. The group based on AIM program in turn could have easily recommended changes in pipe metal from carbon steel to high alloy steel. These observations and recommendations by the RCM team would have helped to avert the failure at Richmond Refinery.

8.5.2 Case study no. 2: Buncefield Terminal explosion [3,4]

The explosion in Buncefield Terminal is primarily attributed to loss of primary and secondary containment. The motor spirit/gasoline (MS) overflowed from the tank in which MS was being received consequently vapor cloud formed and exploded. Fire continued at the terminal for 5 days; 44 people were injured, and the explosion was measured at 2.4 on the Richter scale (Fig. 8.4).

Why did it happen? Were enough layers of protection not envisaged in the design? Or were they dysfunctional and not maintained? The tank was provided with two level gauges with alarm at control room. It also had an independent emergency switch to ensure overfill protection. Sadly, none of them worked.

The tank was fitted with an automatic tank gauging system (ATG) to measure the rising level of fuel and display this on a screen in the

FIGURE 8.4   Image of the fire and explosion: Buncefield Terminal.

control room. But the ATG display "flatlined," for example, it stopped registering the rising level of fuel in the tank although the tank continued to fill. Consequently, the three ATG alarms, the user level, the high level, and the high-high level, could not operate as the tank reading was always below these alarm levels.

The tank was also fitted with an independent high-level switch (IHLS) set at a higher level than the ATG alarms. The IHLS also failed to register the rising level of petrol, so the final alarm did not sound, and the automatic shutdown was not activated. The IHLS did not get activated since the all-important padlock switch was not installed properly. Rather, proper installation of ILHS was not ensured during the project stage and operations too missed to check its efficacy.

A comprehensive AIM program and procedure could have prevented such a catastrophic incident. It is pertinent to mention that tank gauges struck 14 times prior to the incident. Why is it that such a frequent problem was not addressed with all sincerity that it deserves? Is it due to complacency? Or absence of a proper safety management system?

The roles and responsibilities that AIM assigns viz. typically personnel in maintenance, operations, project construction, engineering and safety must together get involved to ensure and certify proper installation of IHLS during project construction including finding solution to frequent stuck-up of level gauge element/improved reliability of the ATG system thereby ensuring safe and smooth operation.

8.5.3 Case study no. 3: Explosion in high-pressure gas line carrying natural gas [5]

Multiple fatality and prolonged production loss in the high-pressure pipeline carrying natural gas (NG) took place at Tatipaka near Andhra Pradesh in India. NG from the well-head station at a pressure of 50 bars is delivered through the gas pipeline. NG is fed to power plants and fertilizers and other industries through long-distance gas pipeline.

The NG drawn from the well contained traces of water and oxides of carbon (CO_x). The wet NG flowing thru the long-distance pipeline caused internal corrosion in the pipeline at the lowest elevation at six o'clock position. The internal corrosion in the line is ascribed to besides others weak carbonic acid corrosion. The NG containing traces of CO_x and moisture formed low pH carbonic acid. Improper pigging of the gas pipeline caused accumulation of contaminants, carbonic acid in the lowest point of the gas pipeline (Fig. 8.5).

The corrosion resulted in a leak in the NG pipeline at the six o' clock position. Each time upon detecting the leak clamping (tightening with a metallic collar around the leaky area of the pipeline to arrest leakage) was resorted to stop the leak until in an overcast morning a profuse NG leak took place. The NG vapor engulfed the surrounding area, and a nearby tea shop provided the ignition source. An explosion followed by massive fire shook the entire area, resulting in multiple fatalities and loss of production.

FIGURE 8.5 Image of the fire and explosion in NG pipeline.

The prime reason for this major incident is attributed to repeat clamping of the high-pressure NG line (an ad hoc measure to stop leakage). Besides this, not installing dryer before NG is fed to the pipeline as conceived in the design; improper pigging of the pipelines; not analyzing the pig residue; failure to close the remote operated sectionalizing valves, etc. contributed to the accident.

Though IPS study for the pipeline was carried out, the same was not properly studied. IPS report indicated substantial thinning of the pipeline which called for urgent replacement of the corroded portion of the pipe section.

Analysis of pig residue was not done in the high-pressure NG pipeline. The analysis of pig residue would have clearly indicated presence of iron in the residue meaning loss of metal indicating continuous corrosion in the pipeline.

Timely closing the remote operated sectionalizing valve would have resulted in minimizing the collateral damage. Provision was built in the design to isolate the sections of pipeline in case of any leak or any other eventuality. But the personnel working in the company could not close the remote operated valve but had to close the valve manually. This resulted in considerable loss of time. NG continued to leak, and fire continued for hours.

Should a high-pressure gas pipeline leaking frequently be subjected to such poor repair and maintenance practices? The organization should have replaced the corroded portion of the pipeline and installed gas dryer to ensure uninterrupted operation. Undertaking regular checks and adhering to the preventive maintenance schedule of various equipment viz. remote operated sectionalizing valves could have reduced the impact of fire and consequential losses.

8.5.4 Case study no. 4: Fire in Pulau Bukom Refinery, Singapore [6,7]

Draining flammable liquid hydrocarbon in open is as good as inviting accident. The risks and hazards associated with it is enormous; more so when the liquid is highly volatile and low in conductivity. Open draining of highly volatile petroleum products heightens the risk multifolds since its vapor will ultimately escape into the atmosphere leading to accumulation of flammable vapor, which in turn may ignite at any point of time.

Fire started in Pulau Bukom Refinery while draining naphtha into open from a pipeline passing thru the pump house. The contractor personnel deployed were involved in draining the naphtha from the pipeline by opening the valves and through an open flange and collected the same into a plastic tray. The activity was undertaken as a part of preparatory work for maintenance.

The collection of naphtha into a plastic tray caused accumulation of static electricity. The frictional force inter alia freefall of highly flammable naphtha into a plastic container generated the static charge, and a spark was good enough to ignite the material. The same phenomenon happened at Pulau Bukom Refinery leading to fire and burn injury to several contract workers.

To extinguish the fire, water was sprayed into the container. The flammable liquid splashed and fire spread affecting the pump house first and subsequently major damage to the property resulting in shutting down of refinery operation.

Identifying risks and eliminating hazards are the prime responsibility of any company in ensuring a safe work environment, particularly for major installations where process safety incidents have the potential for catastrophic consequences. Company failed to address the risks and hazards associated with open draining of naphtha. It failed to undertake job safety analysis and tool box talk viz. educating the contractual workers the risk mitigation measures.

8.6 Human factors

So, what are the human factors in the context of accidents that took place in oil and gas industry? Let us ponder the issues that impede achieving safe, smooth, and reliable operation in installations. As this book looks at process safety from a practitioner's approach and perspective, the author, who has also investigated tens of such incidents, feels that some of the key areas where managers must renew attention include:

- failure to identify the problem areas;
- ascribing to poor maintenance methods and practices;
- overreliance on contractor and consequently poor supervision;
- accepting the assets without proper precommissioning checks;
- too much concern on production and not maintaining the health of assets; and
- inability to assess, evaluate, and analyze the health of assets, and in particular, aging assets, etc.

Why do humans make errors? Human errors are related to the safety culture of the organization. So why errors? Is it due to poor safety culture? Or is it due to absence of safety leadership? Or is it due to undue pressure that leads to stress? Or is it due to lack of motivation? Lack of proper effective training and education?

The competence gap is one of the crucial reasons that can be attributed to some failures. The advent of modern technology, state-of-the-art instrumentation systems, focus on AIM, use of new techniques like RBI and RCM necessitates that proper training and development (T&D) is imparted to the employees to ensure safe and smooth operations. T&D must be an ongoing process, and special attention must be paid to evaluate the learning process. Focus must shift from the numbers of training programs imparted to the quality aspects of training.

The failures leading to major accidents in oil and gas industry that the author has studied can be broadly categorized into two key areas viz. diagnostic error and decision-making error. While diagnostic errors, for example, misinterpretation of an abnormal event are primarily due to lack of competence, expertise, and experience; the decision-making errors occur because there are enough managers in industry who shy away from taking decisions. Indecisive managers continue to doubt their own abilities and such managers quite often end up in wrong decisions.

No manager or supervisor wishes deliberately to create any unsafe situation, but the culture of the organization, to a large extent, makes them ineffective. This is because leaders in the organization "do not walk the talk" or "act as role models." For instance, publicly such leaders advocate that continuous training and development of employees is a prerequisite to good business but, in reality, only those who are "spare-ables" are nominated to attend the training programs. They preach that if you come across any unsafe situation, stop the work, but then pull up the officers for interrupting the operation. Such leaders are insecure; neither have they had trust and confidence in themselves nor do they have trust and confidence in their subordinates. Thus, leaders fail to build teams and the confidence of their followers.

How often have we witnessed organizations celebrating success stories? Celebrating success and honoring the team members in public functions improves motivation and instills feeling of togetherness. Accepting the bright ideas of employees and implementing their ideas enhances the employee engagement process. Employees become part of decision-making process. Where employees work in a team, difficult tasks are accomplished with ease. Doing it right first time becomes habit. Quality gets improved. Errors and violations get reduced. Safety barriers, interlocks, O&M practices improve. Safety becomes the responsibility of the team rather than that of the leaders.

We need leaders to move away from Individualistic approach to Institutional approach; from instructing people to listening to the people and engaging them. We need leaders who prioritize long term process safety assurance over short term budget restrictions and profitability and are not crisis driven. Such leaders will succeed in creating a climate that fosters innovation and creativity in the organization and build a team committed to achieving excellence.

8.7 Way forward and chapter concluding remarks

Organizations willing to extract the benefits of AIM programs must deploy sufficient resources. An effective AIM program enables managers to achieve significant improvement in asset reliability by precisely predicting timely repair and replacement of the assets thus eliminating unexpected failures. Further training the employees on various techniques of AIM significantly improves the efficiency of workforce and quality of the jobs performed.

Application of process safety and elements of PSM facilitates following the O&M practices in the organization in true spirit. To achieve reliable and safe operation, identifying risks and hazards associated with any work is a prerequisite. However, understanding process safety issues requires proper knowledge and in-depth understanding of the subject.

The audit process becomes more meaningful unlike casual site walk-outs. Addressing effectively the gaps pointed out in audits improves safety. Reporting of near-miss incidents and accident−incident investigation improves the learning process. Thus, PSM enhances the overall effectiveness of the organization. People work in tandem and maintain strict vigil on the leading indicators. Silos get broken. Key result areas get measured and accomplished. All these eventually improve the bottom line and productivity of an organization. It is a step toward achieving excellence in all spheres of business.

References

[1] Keim, Kelly Vice-Chair API RP-754 Drafting Committee: ANSI/API RP-754, Process Safety Performance Indicators for the Refining & Petrochemical Industries, U.S. Chemical Safety and Hazard Investigation Board, Houston, TX July 23, 2012, Public Hearing on Process Safety Performance Indicators. <http://www.csb.gov/userfiles/file/keim%20(api)%20-%20powerpoint%20-%20printed.pdf>, 2012 (accessed 31.04.19).
[2] Chemical Safety Board, USA. <https://www.csb.gov/> (accessed 31.04.19).
[3] <www.buncefieldinvestigation.gov.uk> (accessed 31.04.19).
[4] <www.hse.gov.uk/comah/buncefield> (accessed 31.04.19).
[5] <www.csb.gov/dupont-la-porte-toxic-chemical-release>.
[6] <www.mom.gov.sg/shell-fined> (accessed 31.04.19).
[7] <www.gccapitalideas.com/oil-refinery-fire-bukom-island> (accessed 31.04.19).

Management of change

Ram K. Goyal

Bahrain Petroleum Company (BAPCO), Bahrain Refinery,
Southern Governorate, Bahrain

9.1 Introduction

A key development that has enhanced safety in the industry over the past 30 years, as explained earlier in this book, has been the advent and introduction of the process safety management (PSM) systems. A fundamental aspect of a PSM system is the management of change (MOC) element. Most incidents in industry occur during either a changing phase (such as start-up or shutting down) or when a change is introduced to a running system designed to operate in a certain way and within a certain operating envelope, now redesigned, or redefined to operate in another.

This chapter addresses MOC through various case studies and in-depth discussions on their importance and why from the author's experience it is critical to understand the importance of a disciplined approach to creating change and the human elements that come with that change.

9.2 The cases for change

At 8:33 a.m. on June 13, 2013, Scott Thrower (47), an operations supervisor at the Williams Olefins Plant in Geismar, Louisiana, opened a couple of valves to let hot water in the tube side of a propylene fractionator reboiler to bring it online from its standby mode. He opened the heating medium in the belief that the shell-side of the reboiler contained nitrogen. Unbeknownst to him, liquid propane mixture had leaked into the shell-side either through passing block valves or due to a previous mis-operation. Three minutes after he opened the valves, the reboiler shell violently ruptured causing an explosion and fires. Scott lost his life in the incident, and so did a Williams operator, Zach Green (29). The US Chemical Safety Board (CSB) in their investigation report cited shortcomings in the company's MOC as one of the key causes of the incident [1]. The isolation valves on the pair of reboilers were not a part of the original design; and hence the shell sides were protected from overpressure by means of a free path to the pressure relief valves (PRVs) mounted on top of the fractionator column. The company added the isolation valves, utilizing its MOC process, to achieve operational flexibility in that the unit could continue to run with one reboiler online while the other taken offline for cleaning. The MOC process was deficient in that it failed to recognize that single block valves cannot be relied upon to provide leak-light isolation and that the shell-side of the standby reboiler lost its overpressure protection due to blocked path to the PRVs.

The well-known Flixborough disaster in 1974 in the United Kingdom was perhaps the loudest wake-up call for the chemical and petrochemical industry highlighting the need for a thorough review of a change before its implementation [2]. At this Nypro UK facility, a Reactor (No. 5) from a series of 6 had to be taken out of service due a crack noticed on its wall. A hastily fabricated "dog-leg" piping connection was installed to connect Reactors 4 and 6 so that production could be continued, albeit with a reduced efficiency. No drawing of the bypass pipe was made other than in chalk on the workshop floor [3]! The bellows connecting this bypass failed, releasing a large volume of cyclohexane vapor which ignited causing an explosion and fires.

Nonetheless, it took the industry several years—and the Bhopal Tragedy—to develop and implement formal PSM systems which included rigorous MOC systems as well. The American Petroleum Institute issued its Recommended Practice, API RP-750 [4], in 1990, and the US lawmakers promulgated the 29 CFR OSHA 1910.119 regulation [5] in February 1992, which mandated adoption of a formal PSM, and implementation of an MOC system by all process industry in the US became a legal requirement under Paragraph "L" of the regulation.

The primary purpose of this law was to prevent or minimize the consequences of catastrophic releases of toxic, reactive, flammable, or explosive chemicals. Under this law, the employer must establish and implement written procedures to manage changes (except for "replacements in kind") to process chemicals, technology, equipment, and procedures; and changes to facilities that affect a covered process. The procedures must assure that the following considerations are addressed prior to any change:

- the technical basis for the proposed change;
- impact of change on safety and health;

- modifications to operating procedures;
- necessary time period for the change; and
- authorization requirements for the proposed change.

Furthermore, employees involved in operating a process and maintenance and contract employees whose job tasks will be affected by a change in the process must be informed of, and trained in, the change prior to start-up of the process or affected part of the process. If a change results in a change in the process safety information (PSI), then such information must be updated accordingly. If a change results in a change in the operating procedures or practices, then such procedures or practices must be updated accordingly.

9.3 Scope of a management of change

What should be covered under a formal MOC system is not merely a rhetorical question because the word "change" is rather ambiguous and can mean different things to different stakeholders. For example, if a company has embarked on a multimillion-dollar mega-project to expand or upgrade its facilities, then it is surely a "change." However, does this change need to be managed via the MOC system under the PSM umbrella? Perhaps not, since it can be assumed that the overall project management system will be able to address all related PSM, environment, health, safety, security, and corporate social responsibility issues. Likewise, for a major merger and acquisitions event in a company, do we need to invoke the PSM-based management of organizational change (MoOC) system?

The US OSHA 1910.119 was regarded as a law that placed additional responsibilities on the employers which would add to their costs—costs that they were sure to pass on to the consumers of their products and services in the form of raised prices. Hence smaller businesses were excluded from the obligations under this US law; for example, a business that had less than 5 tons of flammable or hazardous materials present at its site was exempt. Nonetheless, all such small businesses will do well to implement PSM principles, not because it is the law but because it makes good business sense as well in terms of long-term survivability and healthy employee relations.

What is "change" and why do we need to manage it? Some Thesaurus entries for "change" are modify, alter, reverse, transform, mutate, exchange, replace, substitute, swap, trade, inverse, transpose, turn, commute, convert, metamorphose, refashion, shift, and so on. In the context of PSM, by "change" we mean any type of change whatsoever. It could be a change in: facilities, plant, equipment, building, processes, technology, operating procedures, process chemicals and catalysts, feed (waste/slop streams injected), process control systems, maintenance plans or procedures, inspection schedules, plant run lengths, personnel schedules, and operator training and qualifications. Note that "change" includes installation as well as removal.

Through the looking glass

"I see you're admiring my little box," the Knight said in a friendly note. "It's my own invention—to keep clothes and sandwiches in. You see, I carry it upside-down so that rain can't get in."

"But the things can get out," Alice gently remarked. "Do you know the lid's open?"

"I didn't know it," the Knight said, a shade of vexation passing over his face. "Then all the things must have fallen out, and the box is no use without them." **Lewis Carroll**

The process starts with a "change originator." Let us say you have an idea and you want to get it implemented in the company (you are the change originator). As the change originator, it is natural for you to ask yourself, "Why do I want this change made?" Is it because it will improve the company's profitability or improve the safety of the plant, or improve the environment or improve the occupational health of the work force or achieve some other desirable outcome? If your answer to all these questions is "No" then why make the change at all? Suppose it is likely that the proposed change will positively contribute to profitability, safety, environment, or health, then why should we not go ahead and make the change? Why do we have to "manage" the change?

Change originators generally fall in love with their ideas. They want them taken on board and implemented right away. This is especially relevant in companies that run some form of good ideas award schemes which result in cash awards or direct incentives to the originators of cost-saving ideas. As change originators, we may have our minds so firmly fixed on the problem we are trying to solve that we sometimes fail to foresee the side effects of our solution. The history of our industry is littered with instances when hastily made changes have led to disasters, the most famous being the explosion at Flixborough in the United Kingdom in 1974, as mentioned earlier. An incident where a "corrective" action taken by a highly dedicated supervisor led to a multimillion-dollar loss is described later in the chapter (Fig. 9.1).

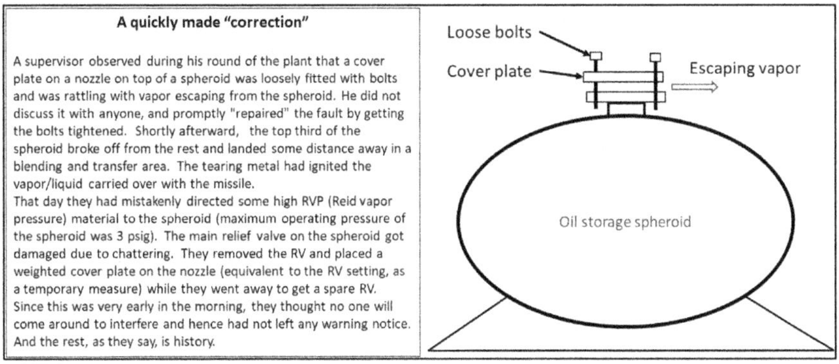

A quickly made "correction"	
A supervisor observed during his round of the plant that a cover plate on a nozzle on top of a spheroid was loosely fitted with bolts and was rattling with vapor escaping from the spheroid. He did not discuss it with anyone, and promptly "repaired" the fault by getting the bolts tightened. Shortly afterward, the top third of the spheroid broke off from the rest and landed some distance away in a blending and transfer area. The tearing metal had ignited the vapor/liquid carried over with the missile. That day they had mistakenly directed some high RVP (Reid vapor pressure) material to the spheroid (maximum operating pressure of the spheroid was 3 psig). The main relief valve on the spheroid got damaged due to chattering. They removed the RV and placed a weighted cover plate on the nozzle (equivalent to the RV setting, as a temporary measure) while they went away to get a spare RV. Since this was very early in the morning, they thought no one will come around to interfere and hence had not left any warning notice. And the rest, as they say, is history.	Loose bolts, Cover plate, Escaping vapor, Oil storage spheroid

FIGURE 9.1   Case study of a quick change.

In citing these major loss events in the context of MOC it was never the intention to imply that change is bad and undesirable. On the contrary, change is essential to the survivability of a company. Your creative and new ideas are the very backbone of continued improvement in our products and services. But change needs to be harnessed and managed properly. In real life, the consequences of a change are a mixed bag. Although the change is expected to result in an improvement in profit, environment, health or safety (or a combination thereof), it may at the same time also have an adverse impact on any one or more of these—albeit to a lesser degree. A change is desirable only if its net impact is positive. And a change costs money (implementation cost). The age-old adage *No good deed shall go unpunished* aptly illustrates the potential for adverse side effects of good ideas. As mentioned earlier, a change is desirable only if its net impact is positive. However, this condition alone is not enough for us to go ahead with its implementation. Since a change costs money, we must also make sure that the change is "cost effective." In a world full of good ideas—all vying for our limited resources—it is essential that we pick only those that bring the greatest benefit per dollar spent. This, in fact, is the cornerstone of a company's risk management system [6].

9.4 Change versus "replacement-in-kind"

If you are replacing a piece of equipment with an identical one or issuing an operating procedure with cosmetic changes only (format, grammatical corrections, etc.), then these do not constitute a change. These are examples of "replacement-in-kind." A word of caution: many a times what might, prima facie, appear to be a case of replacement-in-kind, is not when further details are explored. In some areas such as instrumentation, control, automation, data transmission, and processing newer, faster, and improved versions are becoming available in the market in rapid succession. In these instances, replacement-in-kind is not a viable option and hence these must be pursued under the purview of the company's MOC system.

Most companies have developed and adopted their own MOC systems or have relied upon expert help from consultants and other practitioners. Regardless of how a system is developed, they all must fully comply with the requirements of the local laws as well as adhere to the recognized and generally acceptable good engineering practices and principles of PSM. The primary purpose of a robust MOC system is to ensure that all changes to plant, equipment, building, process, chemicals, and operating procedures are:

1. carefully reviewed prior to implementation.
2. all relevant documentation is updated.
3. all personnel involved are informed and trained. and
4. information about the change is documented.

From a PSM perspective one might try to enhance the robustness of an MOC system by mandating a large set of requirements that proves to be too cumbersome and costly to implement and achieve in practice. Such systems might look good on paper and might favorably impress external auditors but are seldom practiced as intended. The key is to ensure that a modicum of risk-based decision making is permitted in defining the scope of the system (i.e., which changes need to be included in the system) and the designing the extent and depth of the review process required. For example, a guideline could be included to decide on the type of the review deemed sufficient for a proposed change: say a checklist review, or a what—if analysis, or a fully-fledged hazard and operability study (HAZOP), quantitative risk analysis (QRA), environmental and social impact assessment (ESIA), or a combination of these. Generally, the scope of an MOC system includes the following types of changes:

- Changes with a potential impact on environment, health, fire protection, safety, or quality.
- Replacement or addition of equipment other than in kind. (Equipment includes process equipment, piping, machinery, test equipment, safety equipment, control systems, health monitoring equipment, sampling equipment, etc.)
- Permanent removal/dismantling of equipment.
- Operation outside safe, established limits, for example, exceeding maximum allowable operating pressure, temperature, flow, or level— temporarily for a short term, under strict supervision.
- Use of new chemicals, feeds, or catalysts that have not been previously used in the company.

- Introduction of, or modifications and additions to operating, safety, health, or maintenance procedures.
- Temporary repairs and connections.
- Use of new or alternative construction materials.

A carefully designed MOC system will ensure that the number of changes handled in the system per year is workable. Overloading the system with known trivial items can be counterproductive in that these rob time and attention from the more pertinent ones. But then it begs the question, "Who decides what is trivial and what is not?," and we are back to risk-based decision-making. The number of changes generated at a given plant site depends on its age, complexity, and the number of process units located in the plant. In some cases of large plants with aging equipment, the number of changes to be processed in any given year can be in the thousands.

9.5 Management of change review process

Arguably the "review" is the most crucial element in the MOC process. A thorough review takes time and resources but a perfunctorily carried review is meaningless and does not achieve anything. In the aforementioned Williams Geismar plant incident in the beginning of this chapter, any simple what—if review would have alerted the stakeholders to the simple facts that a single block valve cannot be trusted to provide reliable isolation, and that inserting an isolation valve between a process vessel and the pressure safety valve provided for its protection is just asking for trouble.

For process plant related changes possible stakeholders could be from disciplines such as process technology, automation and control, mechanical engineering, operations, maintenance, inspection, EHS, security, electrical engineering, utilities, and civil/structural engineering. Some system might prefer to define various stakeholders in terms of their roles in the MOC process; terms such as change originator, change owner, change owner department manager, actioning department manager, and job owner might be used. Nonetheless, almost all systems define or identify an MOC controller whose primary function include reviewing and approving changes before these are cleared for implementation. In addition to the MOC controller, the job owner is responsible for doing all the work necessary to bring the job to a stage where the change can be implemented; the duties include (but are not limited to):

- oversees the job to completion;
- carries out all the pre-work before preliminary review meeting;
- calls for and chairs review meetings, and ensures minutes are taken;
- follows up on recommendations and actions from review meetings; and

- ensures that all necessary document changes have been made, while also ensuring that all related forms and review meeting minutes have been archived once the change is implemented.

A review process could follow either a "sequential" path or a "parallel" path. In the sequential method, the MOC originator routes the proposal forms and all associated attachments such as drawings, sketches, data sheets, and control logic to the next stakeholder for his/her review and onward routing to others in a sequential fashion. In this case, all subsequent reviewers can also see the comments made by the previous reviewers (Fig. 9.2).

Both approaches have their own pros and cons in terms of effectiveness. The sequential process exerts a certain degree of psychological pressure on the reviewer who holds the process—all stakeholders become aware of the cause of any delays. The adverse outcome of this pressure to meet a given deadline is that some reviewers might find it convenient to carry out merely a cursory review and forward the proposal to the next reviewer in line or recycle it back to the originator.

In the parallel system of review, the reviewers are unaware of the comments and observations made by each other and hence cannot benefit from those. The time taken to complete the review will be governed by the slowest responder. Without holding a joint review meeting of the stakeholders (either face-to-face or video conferencing or a combination thereof), both systems are bound to yield unsatisfactory outcomes from time to time. It is a joint review meeting, coupled with a joint site visit, that provides the foundation for carrying out a structured and formal review. A review team comprising of all key stakeholders reviews the

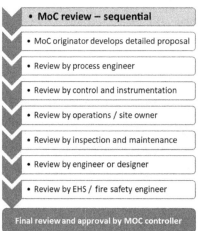
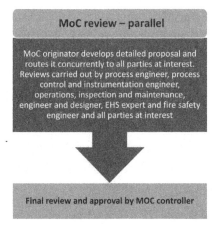
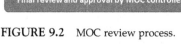

FIGURE 9.2   MOC review process.

TABLE 9.1 Review checklist categories.

MOC team review checklist categories	
Technical basis	Reliability
Environmental considerations	Operability
Health and well-being	Constructability
Fire prevention	Training requirements
Personnel safety	Documents to be updated
Product quality	Drawings and databases to be updated

proposed changes using an MOC checklist or a what—if scenarios list or a formal HAZOP, makes recommendations to minimize any adverse impact of the proposed change, guides the job owner on the need for communicating the change to all parties concerned (including training, if required) and updating of pertinent documents and drawings, makes recommendations for further reviews, and signs off the review meeting minutes. A typical MOC review checklist is expected to contain questions in the following categories, as a minimum (Table 9.1).

Each of the aforementioned categories might contain 10—20 review questions that the team will need to discuss in terms of applicability and compliance [7]. It is important that the team devotes adequate time to questions and issues that matter the most and not digress into areas that are not relevant. At the same time, the team leader needs to ensure that a free-thinking and questioning posture is maintained so that all possible adverse side effects of the proposed change are discussed and adequately resolved.

Effectiveness of a review depends upon how good a team leader is and how good and technically competent the team members are. Given below are a couple of examples in which failures occurred despite technical reviews carried out before implementation of the changes.

In a major US-based oil company, to overcome an environmental compliance problem, they decided to add a small waste stream containing only 40 ppm of organic chlorides, for a short duration, into the feed stream of a process unit. An overhead exchanger in their Naphtha Hydrotreater Plant suffered severe corrosion—the shell wall of 0.500 in. original thickness was reduced to 0.010 in. within a few weeks. Failure of the shell resulted in an explosion and fire. The review team which approved this change had never imagined the extremely rapid loss of wall thickness that occurred.

In a refinery in Africa, during a plant turnaround, an inspector was testing the mechanical integrity of a fractionator column by tapping the inside with a hammer. To his shock and horror, at one place the hammer went through the column wall! You can just imagine the material thickness at that spot. Sometime ago, what appeared to be a minor alteration was made in the operating parameters of this column. The change was approved by stakeholders who reviewed it. The change had resulted in raising the normal operating liquid level in this column. Since the liquid possessed corrosive properties, a portion of the column bottom had been suitably lined with a protective material to prevent corrosion. With the change in the operating parameters, the normal liquid level rose beyond the lined portion! The refinery was extremely lucky to escape what could have been a major disaster.

9.6 Management of change closure

How and when an MOC project can be regarded as completed or given a "closed" status in an MOC administrative system is an important consideration. Its significance has reached new heights in recent times when facility owners face a multitude of internal and external audits and inspections; sometimes separate auditors covering the PSM system, Operational Excellence System, Quality System (ISO-9000), Environmental Management System (ISO-14000), OHSAS (Occupational Health and Safety previously 18001, now 45001), Risk Management (ISO-31000), Government EHS, insurance audits, and company internal auditing processes. Most audit systems consider MOC systems as an integral and significant part of their scope and hence the MOC controller and MOC administrator in a company need to become quite familiar with the demands placed on them by various auditors. Many times the number of people auditing an MOC far exceeds the number of people trying to implement it. This is because "leading indicators," as opposed to "lagging indicators," is the current craze with safety professionals, and parameters generated by an MOC system are considered as valid and effective leading indicators of safety performance of a process unit or a company.

Average time taken to process an MOC and the number of overdue MOC items (change requests that should have been implemented and closed by the stipulated date but have not been) are two parameters that are frequently audited as leading indicators of process safety performance. This generates a great deal of pressure on people involved in implementing the change. The MOC review process can suffer under this pressure to achieve closure within the given deadline. The "Friday Afternoon" (or the last working day of the week) effect takes over especially in the sequential mode of reviews when the MOC controllers might find it tempting to approve all pending MOC items from their inboxes before leaving for their weekends.

Should an MOC be retained in an "open" status even when its further execution and implementation has been transferred to another managed system in the company? Take the example of an MOC that has been properly raised and recorded in the MOC system, alternatives have been reviewed and one has been selected for follow-up, the solution has been designed and then thoroughly reviewed by a team and has been approved. At this stage, the entire project has been transferred to the project management or engineering group of the company who have internal work processes to monitor and chase projects to completion. Should this item be shown as "closed" in the MOC system at the time when it gets transferred to the engineering/project management system or should it be retained in an "open" status in the MOC system?

Rushing to close out an MOC can sometimes lead to situations where the people given the charge to implement it subsequently can make mistakes because they might not be familiar with the entire background and the rationale behind the MOC. This is illustrated by an example incident (Fig. 9.3).

Another leading indicator that is often tracked in audits relates to what can be termed as "temporary changes." The parameter of interest is "how many temporary MOC items have not been removed or dismantled by their stated finish dates." A great many leak repair patches and boxes, temporary sample points, temporary bypasses, temporary drains and vents can become rather permanent fixtures in a plant, creating unnecessary hazards. Hence it is imperative that all MOC closure procedures must be strictly adhered to.

9.7 Management of organizational change

MoOC, or sometimes referred to as organizational change management (OCM), is a term that can include a vast number of topics related to human resource and training management of a company—topics such as corporate standard for OCM, modification of working conditions, personnel turnover, task allocation changes, organizational hierarchy changes, and organizational policy changes [8]. The overall approach to MoOC is no different from that adopted for other type of changes processed under the purview of the PSM system (Fig. 9.4).

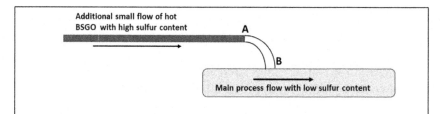

There occurred a need to direct a small stream of hot black slop gas oil (BSGO) to a much larger main feed flow of a process unit. The main feel flow line was carbon steel (CS) and was considered suitable for the low sulfur content and low temperature of the feed. A smaller CS line connected to the main line was available for use for injecting the BSGO stream. It was decided to use this line as a temporary measure till the next plant turnaround when a line of a suitable metallurgy was planned to be installed. It was believed that although CS line was not acceptable in the long run, it was expected to last during the interim period. However, the line leaked after only a few days in service (the sulfur content of the BSGO was significantly greater than what was assumed). The escaping hot BSGO auto-ignited causing a fire in the unit. Luckily, the damage was limited to the immediate vicinity, and no one was injured. An MOC was raised in the system to replace the BSGO line with suitable metallurgy. The MoC was properly reviewed and approved by the review team and all procedures were followed properly. After completion of the design, the project was transferred to the list of "turnaround projects" and was closed from the MOC system. The contractor implementing the MOC as instructed, changed the line and all its components to the required upgraded metallurgy but failed to replace the last elbow (shown as AB in the sketch above). A few weeks after the system was in service, another leak (this time from the last elbow) occurred, followed by a fire!

FIGURE 9.3   Case study: temporary change of service.

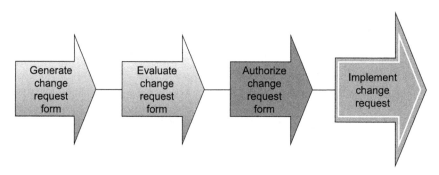

FIGURE 9.4   The change process key steps.

During earlier implementations of PSM systems per 29 CFR OSHA 1910.119, safety professionals were reluctant to include MoOC within the scope of MOC systems partly because the significance of human factors in process accidents was not fully understood. Incidents such as the Phillips Channelview fire and explosion of 1989 and the BP Texas City incident of 2005 amply demonstrated that factors such as personnel training and shift work allocations can become significant contributing causes to major loss incidents in the process industry.

The direct line-of-sight between human resources or manpower planning and process safety can sometimes be rather obscured and unclear.

However, it does not require a great deal of mental gymnastics to see that the following types of "organizational" changes are likely to have direct impact on process safety:

- duration of training specified for operators, maintenance staff, engineers, plant inspectors, and other technical personnel;
- content of training modules;
- frequency and extent of retraining;
- change in roster positions of technical personnel;
- changes in shift personnel or shift times;
- changes in holiday coverage and overtime allocations;
- numbers and/or competency levels of technical personnel; and
- introduction of new technology, chemicals, catalysts, or equipment without adequate training of personnel.

Should MoOC be implemented as a part of the overall PSM-driven MOC system or should it be implemented as a standalone system is a matter to be best decided by the company or the plant site management. As in any such decisions, there are pros and cons to each approach. The subject matter experts and stakeholders that are generally involved in processing a process plant or equipment related MOC will need to involve other experts if MoOC is included in the overall MOC system. Separate review checklists that include questions related to human factors engineering, ergonomics, and occupational health will need to be developed and used.

9.8 "Minor" change

Like any other system or procedure, the successful implementation of a robust MOC system can only be achieved through common sense in interpretation and application of the requirements given in the MOC procedure. The procedure is bound to fail if applied pedantically. It is the "intent" which is important, not the "letter."

To reduce the overall workload associated with a given MOC system, some experts have advocated use of two categories of changes: minor changes and major changes. In practice, this is a rather difficult area to resolve. There have been numerous attempts in other companies to define what constitutes a "small" or "minor" change so that they could specify less stringent review requirements for these. However, in the author's opinion, no one has yet really succeeded. For example, one set of definitions is as follows:

- *Major Change*: a broad scope change that requires input from multiple disciplines and may involve updates of numerous PSI documents.

- *Minor Change*: a limited scope change requiring minimal technical review; a minor change will usually result in updates to one or two specific PSI documents.

These definitions sound reasonable on paper but are extremely difficult to implement in practice. If you get a feeling that sometimes the change is so simple that a formal review seems unnecessary then you have not really grasped the essence of what makes an MOC system robust. The cornerstone of a robust MOC system is that no change is "small." Or, to put it another way, a change may appear "small," but that does not mean its impact is insignificant. Trevor Kletz (a well-known PSM Guru) had illustrated this quite well from several incidents reported in his books [9].

Slight error in metric–imperial units conversion: After a major turnaround-and-inspection of a hydrocarbon product storage tank, a slight error was made, while converting from metric to imperial units, in recording the tank height, the safe operating liquid level, and the emergency shutdown setting that is meant to prevent the tank from overflowing by automatically switching off the pumps feeding the tank. The tank got overfilled; the frangible joint between the shell and the roof parted (as it was designed to do under this scenario) and hydrocarbon was released in the dyke. Fortunately, there was no ignition; the spill was noticed by an outside operator quickly and the spill emergency protocol was initiated.

Slightly longer thermowell tube: Due to a slightly thicker washer being used in a thermowell assembly on a process column, the effective length of the well increased slightly. This resulted in the thermocouple not reaching the contact point and thus not detecting the temperature accurately which caused process upsets and product quality loss. The troubleshooting in this case also presented a difficult challenge because of the unusual nature of the root cause.

Just a different O-ring material: O-rings made of viton were being used in the piping joints which were in process gas service. The process gas contained trace quantities of amine, a fact that was not recognized at the time of selection of the O-ring material. Over a period of time, the presence of amine deteriorates viton O-rings – they turn brittle and then fail in service. A large release of process gas occurred when one such O-ring failed. Luckily, the released gas did not ignite and dispersed without causing a serious impact in terms of safety and health exposure. The overall loss was limited to business interruption caused by unit shutdown for repairs.

An item such as a label on a piece of equipment could be considered just a "minor" thing. It cost a US major oil company a multimillion-dollar fire to learn otherwise—The Sponge Tower Fire [10]. The lesson

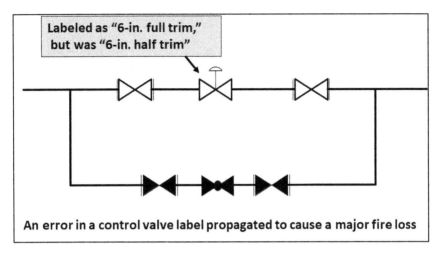

An error in a control valve label propagated to cause a major fire loss

FIGURE 9.5   Case study: assumptions and incorrect labeling.

comes free of charge to us. It is worth describing briefly here what happened (Fig. 9.5).

After a de-bottlenecking project, the flow of sponge oil was controlled by a valve carrying the nameplate "6-in. full trim." However, operations found it necessary to crack open the bypass around the valve to maintain the valve in a control position. The primary reason behind this was that although the valve was labeled as "6-in. full trim," it was actually a "6-in. half trim" valve, thus limiting the maximum flow that could pass through it. Note that "cracking open the bypass valve" for normal operation is an inherently unsafe practice and must not be permitted to become the norm. To resolve this control problem, the valve was replaced by a new 8-in. full trim valve. The designer concluded that if a 6-in. full trim valve (as wrongly labeled) was not large enough to handle the flow then there was a need to increase the size of the control valve to 8-in. full trim. Had a proper MOC review been conducted then it would have come to light that the actual valve was 6-in. half trim and just a replacement to full trim would have sufficed. Nonetheless, in this instant the valve was replaced with an 8-in. full trim size. This worked in operation with no major problems, although the new valve was operating only at 25% valve opening. And then one day, the ambient temperature dropped suddenly. The control system responded to this sudden drop in ambient temperature, but the valve reached a point of instability and began to rapidly open and close. This created a severe surge and major vibration of the piping system causing flange leaks, loss of containment and subsequent ignition, followed by a major fire event.

9.9 Concluding thoughts: modern trends in risk tolerance

Despite implementation of numerous safety systems, procedures and standards, the process industry continues to suffer from major loss events. General public's perception is changing day by day. They fully believe that we, the industry, have the knowledge and the capability to prevent loss events from occurring and that cost is the only thing stopping us from achieving that goal. The public mood is to severely penalize companies that pollute the environment or create any other health hazards. The high cost of clean-up that BP had to endure as a result of the Horizon oil spill is well known. A fire occurred in a San Francisco refinery which emanated smoke for a few hours. Several thousand inhabitants of a local township filed claims against the owner company citing adverse health impact due to smoke inhalation. The insurance companies have become extremely nervous: They had to settle several very high claims resulting from fires and explosions in refineries over almost all parts of the world. Risk tolerance of the public and other stakeholders is wearing thin. The very "license to operate" is on the line!

A robust MOC system can prove to be a valuable tool in preventing loss incidents from your company. Apply it rigorously, deliberately, and intelligently.

A "domino-effect" loss can be prevented even if just one domino stands firm.

References

[1] U.S. Chemical Safety and Hazard Investigation Board (CSB), "Williams Geismar Olefins Plant Reboiler Rupture and Fire, Geismar, Louisiana" Incident Date: June 13, 2013, Two Fatalities, 167 Reported Injuries, No. 2013-03-I-LA.

[2] UK Health and Safety Executive (HSE), The Flixborough Disaster: Report of the Court of Inquiry, HMSO, ISBN 0113610750, 1975.

[3] F.P. Lees, *Loss Prevention in the Process Industries*, Three Volumes, second ed., Butterworth, 1996.

[4] American Petroleum Institute, API RP 750, Management of Process Hazards, first ed., 1990.

[5] Occupational Safety and Health Administration (OSHA): Federal Regulation 29 CFR 1910.119 (February 1992), Process safety management of highly hazardous chemicals.

[6] R.K. Goyal, Practical Examples of CPQRA from the Petrochemical Industries, Trans IChemE, Vol 71, Part B, May 1993 issue, The Institution of Chemical Engineers, Rugby, England, 1993.

[7] American Institute of Chemical Engineers (AIChE), Center for Chemical Process Safety (CCPS), Guidelines for the Management of Change for Process Safety, Wiley, 2008, ISBN: 978-0-470-04309-7, .

[8] American Institute of Chemical Engineers (AIChE), Center for Chemical Process Safety (CCPS), Guidelines for Managing Process Safety Risks During Organizational Change, Wiley, 2013, ISBN: 978-1-1183-7909-7, .

[9] T. Kletz, Still Going Wrong!, Elsevier, 2003.

[10] M&M Protection Consultants, 100 Large Losses - a thirty-year review of property damage losses in the hydrocarbon-chemical industries, 11th/12th/13th editions, Chicago, IL, 1988/90/92.

10

Management of risk through safe work practices

Chitram Lutchman and Ramakrishna Akula

Safety Erudite Inc, Canada

10.1 Introduction

Hazardous industry operations are multifaceted with complex human performance factors. Risks and hazards are more prominent in activities where human intervention is necessary. It is essential to define, communicate, and ensure all operational activities in a hazardous industry are performed in a safe, standardized and consistent manner to manage all kinds of risks.

The chapter starts with a brief discussion on human performance gaps to manage risks in complex process industries and highlights the importance of a safe system of work to manage risks to as low as reasonably practicable levels. It also outlines how to establish a safe system of work in the process industry and provide guidelines to maintain the documentation. Finally, it ends with a discussion on various safety programs to comply with a safe system of work.

10.2 Human factors in risk management

Human behavior and performance are mentioned as causal factors in most accidents on process plants. The human factors in the decision-making process are the main cause of incidents. If the accident rate is to be decreased, the human factor must be better understood, and the knowledge more broadly applied.

Process Safety Management and Human Factors.
DOI: https://doi.org/10.1016/B978-0-12-818109-6.00010-1 **123**

Evidence [1] is drawn from past disasters, such as the incidents at Flixborough, Kegworth, and Moorgate, and the Piper Alpha incident, indicates that failures in, or inappropriate, human behavior were a significant contributory factor.

Human factors study is the scientific discipline connected with the understanding of interactions among humans and other system elements, and the profession that applies theory, principles, data and methods to design to optimize human well-being and overall system performance. The human factors study directly connects with the aspects of behavioral safety and human performance gaps.

10.3 Behavioral safety

Lack of motivation and inconsistent approaches in an organization lead to unsafe acts or poor behavior safety in an organization. Poor safety culture leads to widespread, routine procedural violations, and failure to comply with own companies safety management system (SMS) [2]. Abuse of company SMS enhances risk in operational activities.

It is necessary to address behavioral safety aspects in organizations to manage risks within acceptable limits. For instance, referring to the flow-chart given later, the highlighted results of typical human errors [3], are the concerns in managing risks in operational activities (Fig. 10.1).

10.4 Human performance gaps

Successful operator performance depends on a variety of factors such as vision, leadership, team consensus, individual skills, availability of processes, and tools [4]. Lack of any or combination of these parameters leads to negative performance outputs, as shown in Fig. 10.2.

Anxiety plays a dominant role in actual work execution and the primary factor for any incidents at the workplace. The majority of the operational risks can be addressed adequately if the operator's anxiety levels are brought down to normal. This can be achieved if the operator knows *what* is expected from him in the execution of a specific task or work and *how* to perform correctly. Also, if the assigned task is a routine one, the operator should perform the work in the right manner *every time*. Human performance gaps can be improved grossly by addressing these What, How and Every time in operators' life with the introduction of safe system of work or safe work practices (SWPs).

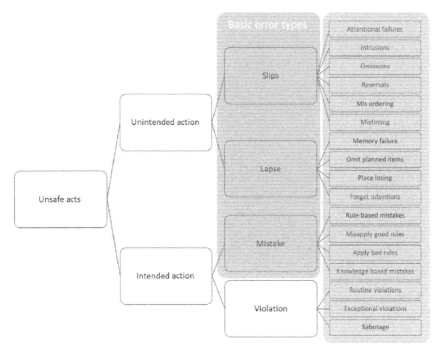

FIGURE 10.1 Risk management concerns resulting from human errors. *Source: Redrawn from A.I. Glendon, S. Clarke, E.F. McKenna, Human Safety and Risk Management, CRC/Taylor & Francis, Boca Raton, FL, 2006.*

Thus, human factors in risk management can be addressed by implementing and efficient monitoring of a safe system of work or SWPs in the process industry.

10.5 Risk management through safe work practices: safe system of work, operating procedures, and safe work practices

A safe system of work is a formal procedure that results from a systematic examination of a task in order to identify all the hazards. It defines safe methods to ensure that hazards are eliminated or risks minimized by defining Operating/maintenance procedures and SWPs.

Operating and maintenance practices are a series of specific steps that guide a worker through a task from beginning to finish in chronological order (www.ihsa.ca). Operating procedures govern activities that generally involve producing a product. Maintenance procedures include typically testing, inspecting, calibrating, maintaining, or repairing equipment.

SWPs are written methods outlining how to perform a task with minimum risk to people, material, environment, equipment, and processes.

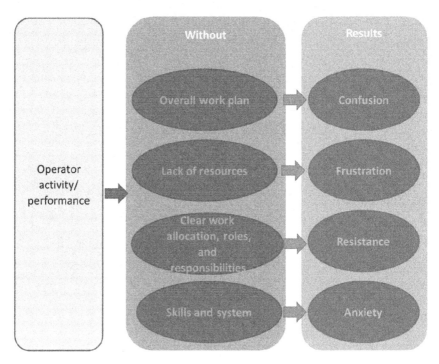

FIGURE 10.2   Factors affecting negative performance. Source: *Safety Erudite Inc., The Operations Excellence Management System Provider, 2019. Retrieved June 02, 2019, from www. safetyerudite.com [5].*

SWPs help control hazards and manage risk associated with nonroutine work and are often supplemented with permits and authorizations to fill the gap between two sets of procedures. Safe work procedures typically control hot work, stored energy (lockout/tagout), opening process vessels or lines, confined space entry, and similar operations (see www. aiche.org).

10.6 Overview: establishing an effective safe system of work

A safe system of work is necessary to establish when hazards cannot be physically eliminated, and some element of risk remains to exist. The principle applies to routine and nonroutine works. The five key steps involved in establishing a safe system are shown in Fig. 10.3.

Particular emphasis to be given to nonroutine activities in a process plant to manage unknown hazards and risks with established SWPs and work authorizations and permits. Thus, the establishment of SWPs starts with defining what is to be done, how to do it in a safe and

Implementation of safe system of work

Assess the task	Identify the hazard	Define safe methods	Implement the system	Monitor the system
• All aspects of task • Consider health and safety hazards	• Identify hazards • Eliminate or reduce risk • Risk rank	• Written procedure • Include required permits/ Authorizations • Specify sequence and safe methods	• Communication • Training • Avoid shorcuts — STOP work Program	• Inspections • Change management

FIGURE 10.3 Key steps involved in the implementation of a safe system of work. Source: *Safety Erudite Inc., The Operations Excellence Management System Provider, 2019. Retrieved June 02, 2019, from www.safetyerudite.com.*

FIGURE 10.4 Concept: Doing the right thing the right way every time. Source: *Safety Erudite Inc., The Operations Excellence Management System Provider, 2019. Retrieved June 02, 2019, from www.safetyerudite.com.*

consistent manner under effective supervision to minimize risks within acceptable limits. In simple words, the operator must know what the right thing to be done, the right way every time which is the cornerstone of operational excellence in terms of operational discipline is.

Fig. 10.4 describes the concept of the right thing, the right way every time [6] in a pictorial manner. The example mentioned in the description is self-explanatory.

10.7 Developing an internal document management system and document framework

Robust internal document management systems contribute to the success of SWPs implementation in any organization. This system should

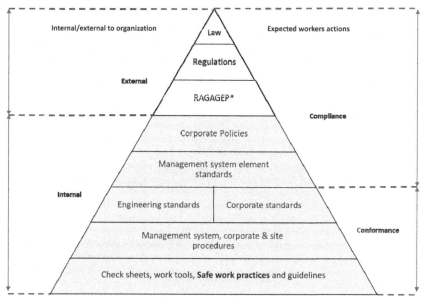

*RAGAGEP — Recognized and generally accepted good engineering practices

FIGURE 10.5 Internal document management system framework. Source: *C. Lutchman, K. Lutchman, R. Akula, C. Lyons and W. Ghanem, Operations Excellence Management System (OEMS): Getting It Right the First Time, 2019, CRC Press/Taylor & Francis Group, Boca Raton, FL [4].*

primarily address the needs of those who want to do actual work without ambiguity. It should be framed with a set of standards, SWPs to execute the work. A typical documentation management structure/SWP support system is shown in Fig. 10.5.

The document framework and internal document management system should address efficient document management controls to monitor revisions and changes. Also, the system should have an efficient auditing system to ensure it is followed consistently throughout the organization. Further, aspects such as roles and responsibilities, training, document storage, and retrieval should be clearly outlined while developing an internal document management system.

10.7.1 Factors to be considered in developing safe work practices

The key factors that must be considered while developing SWPs for nonroutine activities are the system design, job safety analysis, and compliance monitoring program.

10.7.2 System design

System design addresses the development and implementation of SWPs to manage operational risks. Front line leaders must identify the list of tasks to be performed in a specific procedure, record the risks associated with each task, and establish a safe system of work to manage all associated risks in that procedure. Also, the system should clearly define implementation methodologies, and monitoring system such as permit work system, risk assessment, method statements, field supervision to execute the job safely.

10.7.3 Job safety analysis/task analysis

Job safety analysis is another key component in SWPs development. All critical and nonroutine jobs must be analyzed at the task level for associated hazards. To assess a task, first the selected job should be broken down into several simple steps, examine each component for associated risks, and develop measures to mitigate such risks. These mitigation measures must be unambiguous in safe work procedures. The verification or enforcing mechanism should be in place to follow SWPs all times while executing the work with an efficient permit to work system or field supervision.

10.7.4 Compliance monitoring program

SWPs are effectively implemented with compliance monitoring programs. The organizations should have a system to conduct toolbox talks to brief assigned work and to review associated risks. Housekeeping, waste disposal, and emergency response requirements must be communicated before starting the work. Work should be performed under effective supervision and covered under an appropriate permit to work system for safe execution. Depending on the work complexity, periodic supervisory inspections should be conducted, besides field supervision, to ensure the work is executed as per SWPs and that all associated risks are being managed efficiently.

10.7.5 Implementing and communicating safe work practices

Field supervisors play a key role in communicating and implementing SWPs. The field leaders must keep copies of safe work procedures in the field for reference. Front line leaders should periodically review SWPs and all connected procedures and communicate middle management for any modifications or improvements in SWPs as a part of continuous improvements. The middle management must ensure the

SWPs and procedures are current and updated all the time. Also, they should ensure that the procedures and SWPs are aligned with the organization standards and policies. Further, any regulatory requirements or changes in regulatory requirements should update in SWPs promptly and communicate to field supervisors for field implementation and training.

10.7.6 Recordkeeping

Training must be provided to all workers on recordkeeping, such as logs, checklists and records must be verified and signed off during and after completion of the work as per requirements. All permits should be appropriately renewed, signed off, or closed.

10.7.7 User experience writing concepts: user experience writing versus content writing

Conventional document writing is a concern for developing SWPs for field employees, whose education and understanding levels are relatively moderate or low. Generally, field employees are more inclined to practical training than reading conventional documents/procedures to understand a concept. Usually, operators need to search for answers in documents/ SWPs for their questions in the traditional style of writing. SWPs can be made more effective and user-friendly if there are written in a simple and straight forward manner and operators quickly find solutions to their questions in mind. Such simplified documents can be originated with the user experience (UX) writing concept. In this concept, the content is written with cognitive learning techniques, with several bullets and tables to simplify the given information. The margin titles in the document help in reaching the required content quickly and efficiently. The examples given below provide an understanding of traditional writing versus UX writing styles on the same concept (Fig. 10.6).

1. Hazards Identification

EMERGENCY OVERVIEW

This product is extremely toxic and flammable and will be easily ignited by heat, sparks or flames. Explosive mixtures form when vapors mix with air. Vapors may travel to a source of ignition and flash back. May be fatal if inhaled - high concentrations can cause immediate death. Hydrogen Sulfide that is an extremely toxic and flammable gas at low concentrations. Exposures to hydrogen sulfide above 100 ppm are immediately dangerous to life and health (IDLH) and may be fatal. Exposures to hydrogen sulfide between 10 ppm and 100 ppm may produce irritation to the respiratory tract. Refer to North American Emergency Response Guide (NAERG) 117.

FIGURE 10.6 Example of traditional writing style. *Source: Safety Erudite - Usability Mapping training 2019*

Using UX concepts, Flesch-Kincaid Grade Level and Reading Grade levels (RGL), the SWPs can be written to any require grade level to suit operators understanding and educational level (Fig. 10.7).

10.8 Safety programs

Safety programs in organizations should complement the effective implementation of SWPs. Safety *observations or interventions* by a coworker, senior, or supervisor can avoid several unsafe events in the field. *Hazard communications and toolbox talks* provide greater awareness on the task in hand, associated hazards, hazardous material handling, housekeeping, waste disposal, emergency requirements. Management walkabouts ensure alertness in field operators. Also, these walkabouts provide a comprehensive assessment of the field conditions to identify areas for improvement.

Workers' involvement in campaigns and initiatives such as heat stress campaign, stop smoking initiative involvement enhances safety awareness on specific topics and registers in the minds of works for a long time. These program enhancements are designed to overall safety behavior and awareness.

1. Hazard Identification

1.1. Emergency Overview

Properties	The properties are:
	• Extremely toxic and flammable at low concentration
	• Easily ignited by heat, Sparks or flames
	• Explosive mixture forms, when vapors mix with air
	• Vapors may cause flash back
	• Inhalation at high concentration can cause immediate death.

Exposure — Effects of H2S exposure on human body:

Concentration	Effects of Human body
10 ppm to 100 ppm	Irritation to the respiratory tract
Above 100 ppm	Immediately dangerous to life and health (IDLH).

Reference: North American Emergency Response Guide (NAERG-117)

FIGURE 10.7 Example of user experience writing style. *Source: Safety Erudite - Usability Mapping training 2019*

10.9 Summary

Nonroutine operational activities in the process industry are always associated with high-risk ranking due to unknown hazardous associated with it. Human factors such as behavior and performance gaps play a dominant role in managing these risks efficiently. The worker anxiety levels can be brought down with correct and right information on the assigned task, provide necessary tools and supporting information, and ensure work performed under efficient supervision. This can be achieved with a safe system of work that comprises of SWPs and work authorization requirements. Internal document management system and UX concepts enhance the efficient implementation of SWPs in any organization. Finally, a dynamic safety program improves workplace safety and culture.

References

[1] J.W. Stranks, Human Factors and Behavioural Safety, Routledge, Abingdon, Oxon, 2016.
[2] A.I. Glendon, S. Clarke, E.F. McKenna, Human Safety and Risk Management, CRC/Taylor & Francis, Boca Raton, FL, 2006.
[3] J. Reason, Human Error, Cambridge University Press, Cambridge, 2009.
[4] C. Lutchman, K. Lutchman, R. Akula, C. Lyons, W. Ghanem, Operations Excellence Management System (OEMS): Getting It Right the First Time, CRC Press/Taylor & Francis Group, Boca Raton, FL, 2019.
[5] Safety Erudite Inc., The Operations Excellence Management System Provider, 2019. Retrieved June 02, 2019, from www.safetyerudite.com.
[6] C. Lutchman, Process Safety Management Leveraging Networks and Communities of Practice for Continuous Improvement, Taylor & Francis, Boca Raton, FL, 2014.

Process safety information, hazard control, and communication

Azmi Mohd Shariff and Muhammad Athar

Centre of Advanced Process Safety (CAPS), Chemical Engineering Department, Universiti Teknologi PETRONAS, Perak Darul Ridzuan, Malaysia

11.1 Introduction

Due to the rapidly increasing scale and complexity of modern process industries, the mitigation of accidents has become more challenging [1]. These accidents have widespread losses related to human capital, production machinery, company reputation and to the environment. Therefore, the prevention of accidents or reduction of accidents impacts are highly solicited [2]. The accidents during the decade of 1980s, such as Bhopal, India and Piper Alpha, UK, have necessitated the need of developing guidelines/regulations for the industries to comply with for a safer operation. In this context, the guidelines have been formulated by various bodies such as US Occupational Safety and Health Administration (OSHA), Center for Chemical Process Safety (CCPS), Dupont, etc. Moving toward the new laws, OSHA has developed a regulation, Process Safety Management (PSM) of Highly Hazardous Chemicals, 29 CFR 1910.119 in 1992 [3], and for the process industries in the United States, its compliance is mandatory. Similarly, the industries and regulatory bodies worldwide have agreed that PSM would drive a major improvement in process plant safety to protect human and capital assets [4–6]. The implementation of PSM could prevent accidents if process plants follow

Process Safety Management and Human Factors.
DOI: https://doi.org/10.1016/B978-0-12-818109-6.00011-3

the regulation as intended [7]. In this way, with the engagement of OSHA PSM, a significant reduction in number of accidents have been reported, which have ultimately resulted in meeting production targets and enhanced management of process safety and reduction of human error [8].

Currently, the PSM implementation degrees vary from plant to plant due to lacking of systematic technique for industries to comply with PSM requirements and maintain the effective process safety programs [9,10]. Actual implementation costs for the PSM regulation have been orders of magnitude higher than the original estimates [11]. PSM auditing costs are also high, and people are doubtful about its effectiveness [12]. PSM documentation is very difficult and requires great effort. Also, good documentation is just the beginning and proper utilization is again difficult [13]. In general, the PSM implementation requires much effort and time but pays off well if implemented fully [14,15].

The ultimate objective of the PSM is to either avoid or minimize the consequences by the catastrophic leaks of explosive, flammable or toxic chemicals and can lead to various accident scenarios such as fire, toxic release or explosion. As the PSM is specifically applicable for process systems, therefore it is based on all the components needed for a process system. However, the number of PSM elements vary from country to country (and in fact company to company) and its related organization as per their needs. Generally, the PSM elements are divided in four pillars as shown in Fig. 11.1. A comparison of PSM elements in various systems such as OSHA, Environmental Protection Agency (EPA), CCPS risk-based PSM, and the State Administration of Work Safety (SAWS) in China, is provided along with pillars in Table 11.1. It is identified that in all systems, process safety information (PSI) is of vital importance as it manages the compliance data for an industry as well as serves the source of information to manage the hazards. Furthermore, the model presented by Zhao et al. has revealed that PSI serves as the heart of any PSM model and all the elements are connected to PSI [21].

Therefore, PSI needs to be handled with care for better compliance of PSM in any industry. Furthermore, for the accidents in process industries, typically those examples are referred in the literature which have not complied with the PSM standard, however, the examples of companies which have a good PSM program are not referred after the accidents [22]. By studying the all relevant OSHA citation data, it is revealed that incomplete PSI has been the commonly referred [23].

As mentioned earlier, the number of accidents in process plants might reduce significantly after the effective implementation of the PSM standard. However, the investigation of accidents in process industries for the period of 1998 to 2008 by the US Chemical Safety and Hazard Investigation Board has indicated that the decrease in number of accidents is not as per

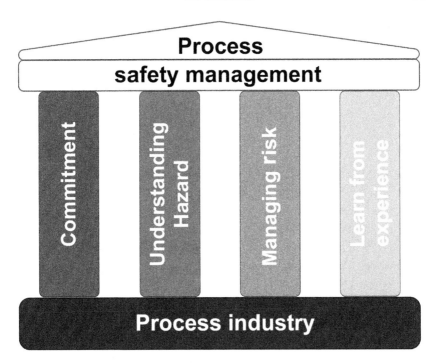

FIGURE 11.1   Pillars for process safety management.

expectations [24]. Among these accidents, 50% of the accidents happened in the industries which have implemented the PSM while the remaining ones occurred in small industries that are not typically covered by PSM [7]. Additionally, the Canadian Chemical Producers' Association (CCPA) has analyzed the 2004 process-related incidents measure analysis (PRIM) report. A total of 89 incidents have been documented and for these incidents, six elements have the 85% contribution [13], as demonstrated in Fig. 11.2. Likewise, a recent analysis for 630 accidents in process industries for the period 1990−2015 has been presented as shown in Fig. 11.3 [25]. The analysis was based on various accident databases such as Chemical Safety and Hazard Investigation Board (CSB), Major Accident Reporting System (MARS), Failure Knowledge Database (FKD) and Zentrale Melde und Auswertestelle für Störfälle und Störungen in verfahrenstechnischen Anlagen (ZEMA). It is observed from this analysis that process hazard analysis and operating procedures are the major contributors in process industry accidents.

From both of these analysis of accidents, it is clearly highlighted that there are potential issues in implementation of PSM. Additionally, it is evident that the PSI has a major role in preventing the accidents. Therefore, it can be understood that for a better compliance of PSM

TABLE 11.1   Comparison of PSM elements in various systems.

PSM pillars	PSM elements			
	OSHA [16,17]	EPA [18]	CCPS RBPSM [19,20]	SAWS [21]
Commitment			Process safety culture	
	Process safety information	Process safety information	Compliance with standards	Process safety information
			Process safety competency	
	Employee participation	Employee participation	Workforce involvement	
			Stakeholder outreach	
Understanding hazard	Process safety information	Process safety information	Process knowledge management	Process safety information
	Process hazard analysis	Process hazard analysis	Hazard identification and risk analysis	Process hazard analysis
Managing risk	Operating procedures	Operating procedures	Operating procedures	Operating procedure
	Hot work permits	Hot work permit	Safe work practices	Work permit
	Mechanical integrity	Mechanical integrity	Asset integrity and reliability	Mechanical integrity
	Contractors	Contractors	Contractor management	Contractor
	Training	Training	Training and performance assurance	Training
	Management of change	Management of change	Management of change	Management of change
	Prestartup safety review	Prestartup review	Operational readiness	Prestartup safety review
			Conduct of operations	
	Emergency planning and Response	Emergency response program	Emergency management	Emergency response
	Trade secrets			

(Continued)

TABLE 11.1 (Continued)

PSM pillars	PSM elements			
	OSHA [16,17]	EPA [18]	CCPS RBPSM [19,20]	SAWS [21]
Learn from experience	Incident investigation	Incident investigation	Incident investigation	Accident management
			Measurement and metrics	
	Compliance audits	Compliance audits	Auditing	Compliance audit
			Management review and continuous Improvement	

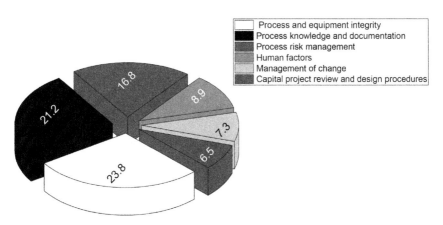

FIGURE 11.2 Contribution of Canadian Society for Chemical Engineering PSM elements in process industry accidents (percentages).

standard in the company, hard work and dedication is needed for the development and management of PSI at all times, as it can contribute toward the hazard control and communication in the company. This chapter briefly discusses this element and related issues specifically the relevant human errors.

11.2 Process safety information element

As mentioned earlier that PSI serves as the heart of PSM system. As per the OSHA standard, all the information regarding the process is

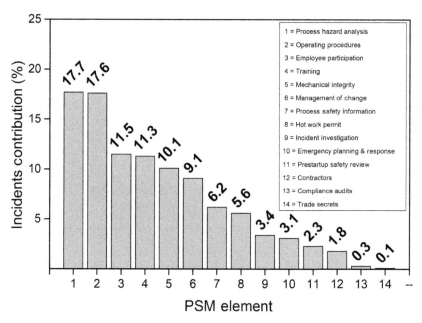

FIGURE 11.3   OSHA PSM elements contribution in process industry accidents.

required for PSM which is provided by PSI. The availability of this information assists both the employees and the employers to know about the hazards associated. Typically, the employees have the expertise in operation, however, the design related information is required in PSM, which is compiled in PSI. The PSI would serve as a resource to multiple users of PSM, such as provides diverse sort of information such as technical information to team conducting the Process Hazards Analysis (PHA) or Prestartup Safety Review (PSSR), the gaps for the generation of better emergency plans for the manufacturing site, and the training programs and operating manuals for the staff. The effectiveness of PSI is merely dependent upon how much accurate and reliable information is compiled [23].

Additionally, for effective PSI implementation, this information must be known and easily accessible to all employees [26]. Besides developing the PSI, managing and updating PSI is also needed such that the information remains updated for all the changes made in the process in terms of chemicals, facility or organizational structure etc. All the changes need to be managed in the system accordingly, such that the hazards and risks related to process could be controlled in an efficient way. Although this target looks simple, however, in the real world, there are many issues in managing PSI such as lack of internal expertise or resources, which may cause major regulatory and safety implications [9,10,13].

Generally, the PSI has three components and information for each component is compiled while developing and maintaining the PSI element. These components include the chemicals being used, technology of the process and the equipment used in the process [17,27]. For the chemicals part, the information source is material data sheets (MDS), which provides information about the characteristics of chemical and its hazardous effects and the remedies for dealing any interaction of chemicals with humans. Furthermore, to maintain the reactivity data of certain chemicals with others is also important, for which a chemical interaction matrix is developed and can be prepared for complete processing plant and for individual sections or process equipment in the facility. An exemplary chemical interaction matrix is demonstrated in Fig. 11.4.

The second component provides the basic information about the chemical process along with the normal and deviated operation scenario and its possible consequences. This information is initially provided by the designer and then maintained by the operating staff such that the latest information is available. In the equipment component, the detailed information about process equipment, safety and electrical equipment are compiled for the deeper information about the process, provided by the designer. The minimum needed information for each component of PSI is depicted in Fig. 11.5, whereas the details of each component is available in the literature [28,29].

11.3 Implementation framework for process safety information

A framework has been proposed by Aziz et al. in numerous works to track and manage the PSI element [30,31]. This framework is based

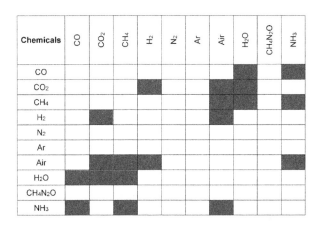

FIGURE 11.4   Example of chemical interaction matrix.

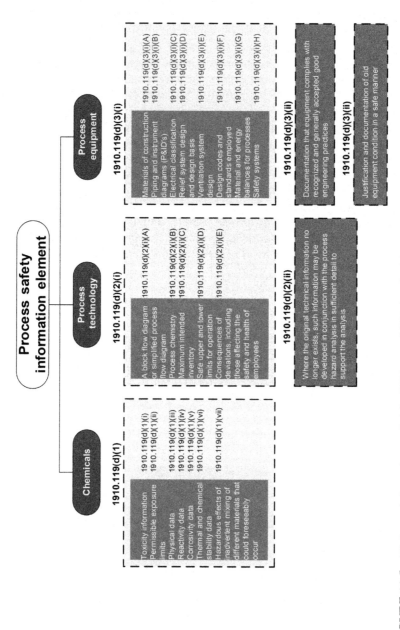

FIGURE 11.5　Minimum information needed for PSI (as per OSHA Standards).

upon the piping and instrumentation diagram (P&ID), as all the details of chemical process are available in this diagram. Furthermore, P&ID is commonly used in process plants for various studies, therefore, the development and management of PSI element using P&ID can be more effective. The framework to engage the P&ID in managing the PSI element is demonstrated in left part of Fig. 11.5. In this framework, the P&ID is divided among several nodes like the hazard and operability (HAZOP) study, which are dependent upon the number of equipment in the process. Then, a node is selected for which several equipment might be available. The PSI implementation for any selected equipment in this node is carried out in accordance to 29 CFR 1910.119(d) as per right-hand part of Fig. 11.6.

In a similar fashion, the information can be compiled and updated for other equipment or streams within the selected node. Once the information for the selected node is done, the data for the next node can be reviewed or updated. This procedure would continue until all nodes in the P&ID and respective equipment are done. The node size can vary which is merely dependent upon the scope of the process it covers. Furthermore, for large process plants, the whole process might not be defined in one P&ID, therefore, similar steps would be repeated for all P&ID to cover the complete process plant. With this technique, the process safety information can be efficiently managed to eradicate the common issues of PSI in managing big data in fuzzy and obscure situations.

11.4 Human errors applicable to process safety information

A number of statistical analysis are available in the literature to identify the major reasons of the accidents in the process industries. It is identified that majorly these accidents have been caused by technical failures [32−34]. A similar conclusion can be drawn for other industries such as nuclear industry [35,36], aviation industry [37], and the other process industries [38]. However, in addition to the technical reasons for the accidents, these are also caused by human and organizational reasons [39]. It is also typical to have all factors contribute to the accidents at the same time [40]. However, for the time being, the differentiation between technical and human errors contributing toward accidents is difficult and needs more research [34].

As both the human errors and the PSI element are contributing in the process industry accidents; there is a link between these two and there could be many human errors relevant to developing and managing the PSI element for a company. The common human errors in process industries are classified among five groups namely: slip, lapses, mistakes which are either rule based or knowledge based and violation [41].

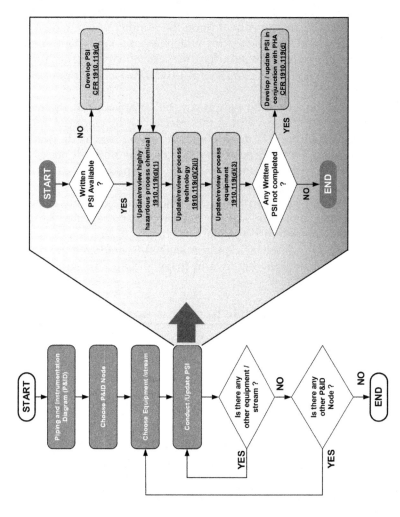

FIGURE 11.6   Framework for process safety information (PSI) and development and management.

Among the various categories of human errors, it is of worth noting that all are applicable to all components of PSI such as chemicals, process equipment and technology. For each component, the possible human errors are discussed underneath.

For the *hazardous chemicals component*, the applicable human errors for PSI are as follows:

1. the information for chemicals is not available in the database;
2. the manual entries of chemical information may lead to communicate wrong information;
3. the data entering person has no expertise about the chemicals and may not give right consideration;
4. the users are not aware about the chemical hazards or the availability of its information in PSI;
5. the change in chemicals or its required concentrations are not updated in PSI;
6. the responsible persons trust themselves about memorizing the changes;
7. the responsible person to update the PSI responds lately to the calls of users for providing information or updating the database; and
8. the work instructions/procedures are not updated regarding the chemicals.

For *process equipment component*, the relevant human errors can be:

1. the detailed information for process equipment is not available in the database;
2. the maintenance record is not updated in PSI and relying upon human memory;
3. the equipment inspection plan is missing;
4. the workers training data is not updated in PSI to work with process equipment; and
5. the specification sheets of process equipment are not available or updated.

For *process technology component*, the related human errors might be:

1. the relevant diagrams such as P&ID, layout diagrams etc. and process description are either not accessible are not updated;
2. the work instructions/procedures are not up-to-date;
3. the information for handling emergencies are not provided; and
4. the workers training and behavior audit is not up-to-date.

These identified human errors can be focused while maintaining the PSI element such that the information regarding the process is up to date in the system and is readily available for better hazard communication and control, ultimately leading to an effective PSM in the company.

11.5 Concluding remarks

The mitigation of accidents in the modern process industries is becoming more challenging, which needs to be catered. In this context, PSM guidelines and regulations have come into existence such that the process industries operate in a safer way. To attain this output from PSM, the compliance to PSM is considered as mandatory in some countries. There are many PSM systems across the globe and regardless of the system, the elements tend to achieve the same objectives. Among these elements, PSI is considered as the backbone for any PSM system because it provides all the necessary information to be needed for managing other elements of PSM. The PSI consists of three components and information regarding each component must be available. These components are process chemicals, process technology, and process equipment.

In this chapter, we have discussed the PSM, compared the elements for various PSM systems across the globe and then the PSI element with its components. For the effective implementation of PSI, a framework using the piping and instrumentation (P&ID) diagram is discussed, which is based upon the OSHA 29 CFR 1910.119(d). As for the effective implementation of PSM, the human effort is of paramount value, therefore, the human errors related to PSI, the backbone of PSM, could have a critical impact on the PSM performance.

Several human errors related to PSI for each component have been identified. These errors, if tackled properly during the management of PSI, would contribute in avoiding the future accidents. Furthermore, the learning from experiences of others in managing the PSI element can also be beneficial to avoid human errors. In such a case, the case studies regarding the experiences of human for PSI element needs to be published to provide more detailed overview of human interaction with PSM. Additionally, much research work is needed to devise the tools to specifically manage the PSI element and generically the PSM system, which can reduce the chances of human error and would be in-line with what is currently known as the industry 4.0 theme.

Acknowledgment

The authors gratefully acknowledge the funding assistance and the facilities of the Universiti Teknologi PETRONAS for this work.

References

[1] G.L.L. Reniers, B.J.M. Ale, W. Dullaert, B. Foubert, Decision support systems for major accident prevention in the chemical process industry: a developers' survey, J. Loss Prev. Process. Industries 19 (6) (2006) 604−620.

[2] M. Demichela, N. Piccinini, How the management aspects can affect the results of the QRA, J. Loss Prev. Process. Industries 19 (1) (2006) 70−77.

[3] A. Mohd Shariff, H. Abdul Aziz, N.D. Abdul Majid, Way forward in Process Safety Management (PSM) for effective implementation in process industries, Curr. Opin. Chem. Eng. 14 (2016) 56−60.

[4] J.F. Louvar, How to prevent process accidents, Process. Saf. Prog. 30 (2) (2011) 188−190.

[5] A. Mohd Shariff, H. Abdul Aziz, M.R. Roslan, K.H. Yew, Protecting Live and Business Through Best Process Safety Management Practices, Proceedings of Asia Pacific HSE Forum in Oil & Gas & Petrochemicals, Fleming Gulf Conferences, Kuala Lumpur Malaysia, 2011.

[6] R. Pitblado, Global process industry initiatives to reduce major accident hazards, J. Loss Prev. Process. Industries 24 (1) (2011) 57−62.

[7] J.F. Louvar, Improving the effectiveness of process safety management in small companies, Process. Saf. Prog. 27 (4) (2008) 280−283.

[8] D.A. Crowl, J.F. Louvar, Chemical Process Safety: Fundamentals with Applications, Pearson Education, 2001.

[9] W.F. Early, Database management systems for process safety, J. Hazard. Mater. 130 (1) (2006) 53−57.

[10] M. Kho, Strengthening process safety requirements in Hse management system an Noc's experience, in: AIChE PPSS Conference, Ernest N. Morial Convention Center, 2008.

[11] W.G. Bridges, The cost and benefits of process safety management: Industry survey results, Process. Saf. Prog. 13 (1) (1994) 23−29.

[12] S. Arendt, Continuously improving PSM effectiveness—a practical roadmap, Process. Saf. Prog. 25 (2) (2006) 86−93.

[13] P.R. Amyotte, A.U. Goraya, D.C. Hendershot, F.I. Khan, Incorporation of inherent safety principles in process safety management, Process. Saf. Prog. 26 (4) (2007) 333−346.

[14] H.A. Aziz, A.M. Shariff, M.R. Roslan, Managing process hazards in lab-scale pilot plant for safe operation, Am. J. Eng. Appl. Sci. 5 (1) (2012) 84−88.

[15] M.I. Rashid, N. Ramzan, T. Iqbal, S. Yasin, S. Yousaf, Implementation issues of PSM in a fertilizer plant: an operations engineer's point of view, Process. Saf. Prog. 32 (1) (2013) 59−65.

[16] OSHA (Occupational Safety acbend Health Administration), Process Safety Management. OSHA 3132. US Department of Labor, Occupational Safety and Health Administration, Washington, DC, 2000.

[17] OSHA (Occupational Safety acbend Health Administration), Process Safety Management of Highly Hazardous Chemicals. 29 CFR 1910.119. Available from: <https://www.osha.gov/pls/oshaweb/owadisp.show_document?p_table = STANDARDS&p_id = 9760>, 2013 (last accessed 16.05.19).

[18] EPA (Environmental Protection Agency), Chemical Accident Prevention Provisions. 40 CFR 68. Available from: <https://www.epa.gov/rmp/general-rmp-guidance-appendix-40-cfr-68>, 2004 (last accessed 16.05.19).

[19] CCPS (Center for Chemical Process Safety), Guidelines for Risk Based Process Safety, John Wiley & Sons, 2007.

[20] CCPS (Center for Chemical Process Safety), Introduction to Process Safety for Undergraduates and Engineers, Wiley, 2016.

[21] J. Zhao, J. Suikkanen, M. Wood, Lessons learned for process safety management in China, J. Loss Prev. Process. Industries 29 (2014) 170–176.

[22] W.B.L. Mostia, Got a risk reduction strategy? J. Loss Prev. Process. Industries 22 (6) (2009) 778–782.

[23] I. Sutton, Process Risk and Reliability Management: Operational Integrity Management, Elsevier Science, 2010.

[24] M. Kaszniak, Oversights and omissions in process hazard analyses: lessons learned from CSB investigations, Process. Saf. Prog. 29 (3) (2010) 264–269.

[25] P.H. Siong, K. Chin, H. Bakar, C. Ling, K. Kidam, M. Ali, et al., The contribution of management of change to process safety accident in the chemical process industry, Chem. Eng. Trans. 56 (2017) 1363–1368.

[26] C. Reese, B. Taylor, Surviving and thriving in the era of enhanced OSHA PSM audits, Hydrocarbon Process. 91 (11) (2012) 47–48.

[27] E. Mason, Elements of process safety management: part 1, Chem. Health Saf. 8 (4) (2001) 22–24.

[28] M.S. Dennison, OSHA and EPA Process Safety Management Requirements: A Practical Guide for Compliance, Van Nostrand Reinhold, 1994.

[29] K. Balan, Practical Process Safety Management, Prism Consultants, 2016.

[30] H.A. Aziz, A.M. Shariff, R. Rusli, Managing process safety information based on process safety management requirements, Process. Saf. Prog. 33 (1) (2014) 41–48.

[31] H.A. Aziz, A.M. Shariff, R. Rusli, K.H. Yew, Managing process chemicals, technology and equipment information for pilot plant based on Process Safety Management standard, Process. Saf. Environ. Prot. 92 (5) (2014) 423–429.

[32] K. Kidam, M. Hurme, M.H. Hassim, Technical analysis of accident in chemical process industry and lessons learnt, Chem. Eng. Trans. 19 (2010) 451–456.

[33] F. Lees, Lees' Loss Prevention in the Process Industries: Hazard Identification, Assessment and Control, Elsevier Science, 2012.

[34] M. Athar, A. Mohd Shariff, A. Buang, M. Shuaib Shaikh, M. Ishaq Khan, Review of process industry accidents analysis towards safety system improvement and sustainable process design, Chem. Eng. Technol. 42 (3) (2019) 524–538.

[35] J. Park, W. Jung, J. Ha, Y. Shin, Analysis of operators' performance under emergencies using a training simulator of the nuclear power plant, Reliab. Eng. & Syst. Saf. 83 (2) (2004) 179–186.

[36] J.K. Vaurio, Human factors, human reliability and risk assessment in license renewal of a nuclear power plant, Reliab. Eng. Syst. Saf. 94 (11) (2009) 1818–1826.

[37] K.L. McFadden, E.R. Towell, Aviation human factors: a framework for the new millennium, J. Air Transp. Manag. 5 (4) (1999) 177–184.

[38] S.H. Yang, L. Yang, C.H. He, Improve safety of industrial processes using dynamic operator training simulators, Process. Saf. Environ. Prot. 79 (6) (2001) 329–338.

[39] K. Kidam, M. Hurme, Statistical analysis of contributors to chemical process accidents, Chem. Eng. Technol. 36 (1) (2013) 167–176.

[40] K. Kidam, M. Hurme, Analysis of equipment failures as contributors to chemical process accidents, Process. Saf. Environ. Prot. 91 (1) (2013) 61–78.

[41] P. Kumar, S. Gupta, M. Agarwal, U. Singh, Categorization and standardization of accidental risk-criticality levels of human error to develop risk and safety management policy, Saf. Sci. 85 (2016) 88–98.

Prestart-up and shutdown safety reviews

Chitram Lutchman and Ramakrishna Akula

Safety Erudite Inc., Canada

12.1 Introduction

In this chapter, the authors address aspects of prestart-up and shutdown safety reviews (PSSR) as a final verification that all process safety requirements are addressed as for the start-up of assets in the following situations:

- When new or modified equipment is introduced;
- After turnarounds and major maintenance work; and
- Following the implementation of a major change in processes, the introduction of new materials, or installation of new equipment.

PSSRs are generally performed by multidisciplinary teams whose review of all process safety requirements is intended to verify and assure conditions are such that it permits the safe start-up of a facility or assets. The multidisciplinary teams generally comprise of subject matter experts (SMEs) from the following disciplines as applicable: operations, technical, design, maintenance, health, safety, and environment (HSE) representatives.

PSSR focuses on the following:

- Quality assurance that ensures process equipment is fabricated in accordance with design specifications, and assembled and installed properly.
- Operations readiness that ensures procedures are developed, personnel are trained, and competent and corrective actions are addressed before start-up.

Process Safety Management and Human Factors.
DOI: https://doi.org/10.1016/B978-0-12-818109-6.00012-5

PSSRs are process safety requirements designed to ensure the safe start-up of assets.

12.2 Why is a prestart-up safety review required?

PSSRs provide a final check of new or modified equipment, assets, and facilities to confirm that all appropriate requirements of process safety have been addressed satisfactorily so that the equipment, asset, or facility can be started up and operated safely and without incidents.

All process or manufacturing operation is subject to the following on an ongoing basis:

- wear and tear;
- continuous improvements;
- replacement of equipment in kind (*like for like*) and with new equipment;
- technological changes;
- expansion and mothballing of equipment and assets;
- changes in people in critical roles; and
- process inputs and chemical changes.

When these conditions occur, maintenance, repairs, and change will occur to ensure assets continue to operate safely, and the business can take advantage of economic opportunities.

Caudill [1] advised that a robust application of process safety requirements with preventative, predictive attributes can reduce the occurrence of incidents. To verify the application of these attributes, PSSR provides a reliable process for doing so.

12.3 Prestart-up safety review considerations

When PSSRs are required, core considerations shall include the following:

Considerations	Requirements
Technological changes	Technology is appropriately selected and documented as follows: • Process safety information (i.e., hazards of materials, asset design basis, and process design basis) has been appropriately documented, communicated end-users • Information is readily available to facility personnel as and when they need it

	• Process hazard analysis corrective actions have been implemented, and appropriate actions associated with start-up have been completed
Documentation: Operating procedures, safe work practices and	Operating procedures and safe work practices: • Have been written, authorized, and are adequate • Are consistent with the process safety information • Incorporated appropriate process hazard analysis corrective actions • Procedures are consistent with process safety information and preincident plans
Management of change	• Changes were properly authorized by competent personnel and recorded • Field changes made during project construction should be checked for proper authorization and recording • Personnel changes at all levels should satisfy criteria that establish minimum levels of knowledge and experience • This is particularly important immediately following the start-up of a large project • A system is available to manage subtle and non-engineered changes
Personnel	• Personnel have received or will receive training on specific safety and health hazards, procedures, operating fundamentals, and emergency response • Contractor safety is addressed, they know of any potential fire, explosion, or toxic-release hazards at or near their place of work
Emergency responses	• Procedures are in place to investigate incidents and implement corrective actions • A documented emergency response plan is in place, and personnel are trained in how to respond
Facilities	• Quality assurance procedures have been followed to fabricate process safety-critical equipment in accordance with design specifications and to assemble and install it properly • Systems tests and inspections of critical equipment has been established: • Reliability engineering analysis • Maintenance procedures • Safety interlock checks
Follow-up	• Following start-up, the new facility should be included in the process safety management audit system

12.4 Key roles and responsibilities in prestart-up safety reviews

Key roles and responsibilities involved in PSSRs are detailed in the following table:

Role	Key responsibilities
Corporate Leaders	Corporate leaders are responsible for: • Establishing corporate expectations regarding process safety and PSSR • Developing and approving written standards for PSSR and its application across the organization • Auditing the business to verify the adequate and appropriate application of process safety and PSSR requirements
Business Units and Functional Leaders	Business Unit and Functional Leaders are responsible for the following: • Demonstrated leadership and commitment to process safety and PSSR • Establishing written procedures for commissioning new or modified assets • Providing sufficient competent resources to implement PSSR requirements • Perform periodic self-assessments to assess compliance to PSSR and its application effectiveness • Creating multidisciplinary PSSR teams with the appropriate range of skills and experience to adequately assess start-up readiness • Tracking and implementing corrective actions from PSSRs; verifying corrective actions are completed
Site and Project Team Leaders	Site and project team leaders shall participate in PSSRs for assets they have engineered, procured or constructed
Site Leader or Designate	Site leader or designate is responsible for the following: • Verifying PSSRs are completed to required standards • Verifying corrective actions are closed before startups • Approving PSSRs when they are performed

12.5 Prestart-up safety review team

The PSSR team's objective is to conclude that the equipment is safe to start-up after completion of corrective actions required before the start-up. The composition of the PSSR team is the key to maximizing its effectiveness. It should be multidisciplinary and include expertise that is capable of operating and maintaining the asset. Other expertise or specialists may be included on an as-needed basis (e.g., electrical, instrumentation and control, ergonomics, and software). Consideration should be given to having one or more safety and occupational health expertise on the PSSR team for all PSSRs.

The team membership shall be established commensurate with the hazards, risk and magnitude of the impact of the change and shall include representatives as appropriate from

- Operations;
- Technical groups;
- Maintenance and reliability; and
- Environment, health, and safety.

Prior to the introduction of hazardous materials or energy to new or modified assets, the multidisciplinary PSSR team shall be implemented to verify the following:

- Process hazards mitigation recommendations required before start-up are complete
- Equipment and process testing and inspections are complete
- General requirements for health, safety, and environmental considerations are adequate and will include the following factors:
  - Ergonomics
  - Fire protection
  - Loss of primary containment
  - Environment
  - Physical conditions of assets, process connections and supporting infrastructures
- Operating and maintenance procedures are documented and approved
- Workers are trained, competency assured and competent
- Adequate supervision is available and provided
- Construction and equipment are in accordance with design specifications
- Other process safety management (PSM) elements and requirements are addressed to ensure safe and reliable startups and sustainment
- A field physical inspection is conducted that is supported by documented and relevant stakeholder checklists

12.6 Prestart-up safety review team composition

Generally, team composition guidance is as follows.

Type	Team requirements
For process safety-critical assets	• The team should consist of three or four representatives from different stakeholder disciplines (e.g., manufacturing, maintenance, design and technical, and appropriate safety representatives) • The diverse composition promotes a more robust and valuable review

	• People who operate or maintain the facility (e.g., an operator and a mechanic) should also be included
Introduction of new hazards (i.e., process, material, or environmental)	• The PSSR team should include external support with relevant knowledge and experience in these hazards • These could be manufacturing, maintenance, technical or appropriate HSE representatives
For significant capital projects and for larger projects	• The PSSR may morph into a formal PSM audit • Leaders should ensure a team with the skills and capability to conduct a PSM audit is included

12.7 Prestart-up safety review team leader

The leader of the PSSR team should foster an open atmosphere where the project team members consider the review as an opportunity to ensure a safe and reliable start-up is achieved. The review should not be seen as a referendum on the quality of work nor the competence of the project team or organization. Failure to create this environment will generally result in some project and start-up personnel becoming defensive with a resultant ineffective or compromised PSSR.

12.8 Process safety management assessments

A PSM element assessment is recommended for any asset. PSM assessments can be conducted by using either a PSM checklist or a previously applied PSSR checklist.

Decisions on the scale and timing of the PSM assessment are the responsibility of the BU/facility leader. Where the PSSR is to be done on assets involving significant capital expenditures and one or more complex unit operations with hazardous materials and activities, it is recommended that the PSM element assessment be conducted by using a PSM audit checklist. In such cases, it may be helpful to carry out a two-part review.

1. The first should take place when the design is complete, and construction is at an early stage or has not yet begun.
   a. This allows operating procedures, emergency planning, and training to be upgraded well in advance of start-up.

2. The second part of the review should take place just prior to the introduction of hazardous substances and focus on checking that construction and equipment are in accordance with the design.

For smaller projects or modifications, the PSM element assessment can be conducted using a PSSR checklist. The PSSR team should select appropriate, relevant parts of a PSSR checklist to review for the project or modification.

12.9 Implementing a prestart-up safety review

An effective PSSR is the outcome of a planned and carefully implemented and executed the event. A PSSR generally constitutes of three stages as follows:

Stage	Activities
Planning	• Generate and approve PSSR plan
Implement	• Execute PSSR • Approve PSSR report and action plans
Action management	• Implement prestart-up corrective actions • Undertake start-up • Implement poststart-up corrective actions

12.10 Generating and approval of the prestart-up safety review plan

The PSSR plan should typically include the following considerations:

Considerations	Details
PSSR methodologies to be used	The method chosen should be documented, such as prewritten checklists for physical inspections. A physical field inspection must be guided by prewritten checklists as per the organization's PSSR standard. PSSR methodologies may include one or more of the following: • Physical inspections of the asset • Document reviews • Interviews • Demonstrations and/or observations
Team Composition	The composition of the PSSR team should be cross-functional and, in some cases, may include cross-business area SMEs. • Members should be identified with the specific areas of the plan he/she is responsible for evaluating based on SME capabilities.

	• All teams should include, at a minimum, people who will operate and maintain the asset and the technical people who understand the change.
	• Other personnel or specialists may be included on an as-needed basis (e.g., electrical, instrumentation and control, ergonomics, process and software specialists).
	• Consideration should be given to having one or more regulatory, environmental, and/or health and safety professionals on the team.
	• For minor changes, the PSSR team may consist of only two people.
Schedule	The schedule should address the timing of the final and any phased PSSRs.
	• Personnel identified with conducting the PSSR should be assigned as early as practical prior to actual implementation.
	• This allows for an opportunity for personnel to input their expertise in the development of the plan and to be trained in assessment techniques.
Timing of Implementation	Careful consideration should be made to determine when to execute the PSSR.
	• The risk of conducting the PSSR too early is that it may be ineffective and/or result in too many corrective actions.
	• Conversely, conducting the PSSR too late can adversely affect the quality of the PSSR and/or the start-up schedule.
Cost	Identify the cost based on the scope and resource requirements of the PSSR.
Action management	The plan to manage and track the prestart-up and poststart-up corrective actions.
	• Prestart-up corrective actions
	○ Can result in major incidents and significant impact on production
	○ Can only be completed prior to start-up
	○ Are required by regulations
	• Poststart-up corrective actions: generally does not affect the integrity of the asset nor does it significantly impact production
Approval	Prior to approval, the PSSR plan should be reviewed to ensure it is appropriate for the scope and complexity of the change, and that resources are appropriately allocated.

12.11 Executing the prestart-up safety review

PSSRs are designed to obtain an accurate assessment of the status of change so that start-up and operation can occur without incidents and consistent with requirements and expectations. It is also designed to provide corrective actions that address the findings from the PSSR. The timing of the PSSR is also important in managing risk.

When conducting physical inspections, it is best that PSSRs be executed as a team effort and carried out jointly with all members of the PSSR team.

Ideally, a PSSR is performed jointly and at the same time by the multidisciplinary team performing a system walk-down and verification that the asset meets the quality assurance and operations readiness requirements for safe start-up and operations of the asset.

- Document reviews and verification may be performed independently and sequentially by relevant stakeholder groups and disciplines.
- A lesser acceptable approach is for stakeholder groups to perform independent system walk-downs and reviews when schedule conflicts and availability of SMEs prevent a performing the PSSR at the same time.
- The same PSSR checklist or relevant SME section must be updated in the PSSR report.

12.12 Field and physical inspections

The physical inspection during a PSSR should employ a checklist (see Appendix) that includes the following:

Aspect	Details

A title page	• Identifies equipment has been inspected, when and by whom, including:
	   ◦ A statement that the installation is consistent with design specifications
	   ◦ A statement that the PSSR team concludes that the facility is safe to start-up after certain corrective actions have been satisfactorily resolved
	   ◦ A record of corrective actions (if any) with timing and responsibility for completion either before or after start-up

Occupational Health and Safety	A check that the following basic safety and occupational health areas have been appropriately addressed: • General safety • Machinery safety • Ergonomics • Occupational health • Industrial hygiene
PSM Element requirements met	Verification that the following PSM elements and topics have been appropriately addressed: • Process safety information (e.g., hazards, process design, asset design basis, and Process safety-critical equipment identification) • Management of change - engineered and non-engineered • Process hazard analysis (e.g., hazards, corrective actions, and communication of results)

- Quality assurance (e.g., specifications, vendor inspections, certificates, and installation inspections)
- Mechanical integrity (e.g., documented maintenance procedures, training, inspection frequencies, and spare parts listing)
- Operating procedures and safe work practices
- Training and competency assurance
- Contractor safety management
- Safety instrumented systems
- Regulatory compliance

- A check that other relevant environments, health and safety topics have been addressed including:
  - Community awareness and emergency response
  - Electrical safety
  - Fire protection

12.13 Approval of the prestart-up safety review report and corrective actions

The PSSR report review and approval of the report should verify the following:

- Construction and equipment are in accordance with design specifications.
- All elements of process safety have been addressed, as appropriate.
- Safety, health, environmental, and fire protection items have been checked during a physical inspection of the facility.
- People are trained and competent to start-up and operate the new asset after the change has been accepted.
- Prestart-up corrective actions have been assigned and tracked to completion.

Following the review of the report and assigning of poststart-up corrective actions, signatories on the PSSR are communicating that it is safe to start-up the asset. Approval for start-up should only be granted by members of the team upon verification of completion of action items required for the safe and reliable start-up of the asset.

12.14 Corrective action management

Action items required prior to start-up are actions are those that

- can only be completed prior to start-up;
- are required by regulations;

- prevent potential loss or adverse impact on process safety, personnel safety, the environment, or production.

All actions, including those that can be deferred until after start-up, shall be documented and tracked to completion. A report of open corrective actions should be published and reviewed periodically. The report should highlight any corrective actions that are past the due date or have been extended.

12.15 Prestart-up safety review completion and closure

The closure of a PSSR should comply with the closure and completion of the Management of Change (MOC) requirements. The completed PSSR report should be stored with the MOC documentation as follows:

- For the life of the asset.
- Consistent with the records management requirements of the organization.
- Based on regulatory requirements associated with the change.

12.16 Conclusion

PSSRs, when completed properly, is designed to prevent major incidents from occurring during start-up and subsequent operations of the asset. Leveraging the team effort of a carefully selected group of SME, PSSRs is among the most effective ways for eliminating workplace incidents when changes are made to existing and operating assets.

Reference

[1] J. Caudill, Support robust process safety management platforms with less reliance on incident investigations, Hydrocarbon Process. 98 (7) (2019) 71–73. Retrieved November 10, 2019 from EBSCOHost database.

13

Contractor management

Chitram Lutchman and Ramakrishna Akula

Safety Erudite Inc., Canada

13.1 Introduction

In this chapter, the authors address aspects relating to effective contractor management, which will include contractor prequalification, selection, performance monitoring and assessment, and contract closeout. The authors explore the importance of understanding the business integration processes and human interface when using contractors and how to overcome the challenges faced in the outsourcing of services. This chapter will also explore best practices in improving partnerships between contractors and owners.

13.2 Overview of contract life cycle

Today, organizations are increasingly leveraging outsourcing and contracting of work to third-party materials and service providers. In all instances, the procurement of contracted services and support follows a process defined in the simplified contract life cycle, shown in Fig. 13.1.

The contract life cycle provides a structured approach to the selection and use of contractors in the business and is designed to optimize performance derived from contractors and service providers.

Detailed steps in the contractor, health, safety, and environment (HSE) management life cycle, is provided in Fig. 13.2.

Process Safety Management and Human Factors.
DOI: https://doi.org/10.1016/B978-0-12-818109-6.00013-7
159

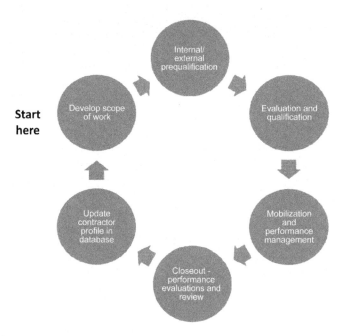

FIGURE 13.1   Contract life cycle. Source: *Safety Erudite Inc. (2019).*

13.3 Reasons for contracting work

According to Keay [1], reasons for contracting services are buried in the following reasons:

- Contractors provide cold eye reviews and a fresh set of eyes: Contractors may be brought in for short periods to provide subject matter expertise (SME) in identifying deficiencies, opportunities, and corrective actions implementation to improve business performance.
- To fill a short-term need for additional and immediate resources to augment existing resources during unexpectedly busy periods and short-term voids.
- To facilitate the delivery of special projects and support for the implementation of new systems. Typical of process industry to support turnarounds and major projects development.
- To provide support for rapid development and growth through mergers and acquisitions and beyond organic growth.
- Contracted work provides test running opportunities for procured services before permanently filling roles and opportunities in the organization.

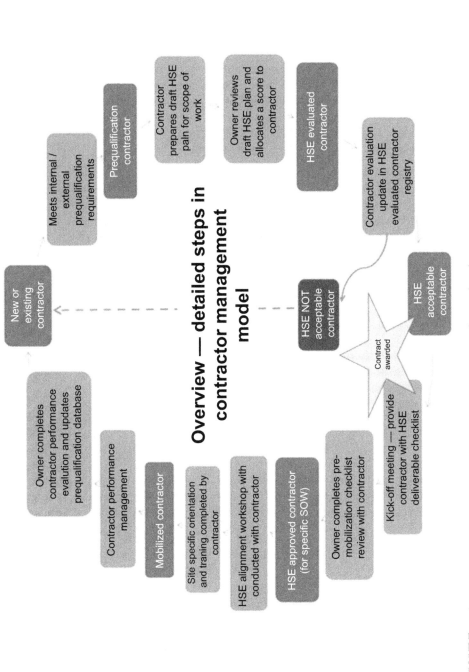

Overview — detailed steps in contractor management model

- New or existing contractor
- Meets internal / external prequalification requirements
- Prequalification contractor
- Contractor prepares draft HSE paln for scope of work
- Owner reviews draft HSE plan and allocates a score to contractor
- HSE evaluated contractor
- Contractor evaluation update in HSE evaluated contractor registry
- HSE NOT acceptable contractor
- HSE acceptable contractor
- Contract awarded
- Kick-off meeting — provide contractor with HSE deliverable checklist
- Owner completes pre-mobilization checklist review with contractor
- HSE approved contractor (for specific SOW)
- HSE alignment workshop with conducted with contractor
- Site specific orientation and traning completed by contractor
- Mobilized contractor
- Contractor performance management
- Owner completes contractor performance evalution and updates prequalification database

FIGURE 13.2 Contract HSE management life cycle. Source: *Safety Erudite Inc. (2019)*

Generally, however, where contractor management and contractor safety management becomes relevant is primarily in process, manufacturing and construction industries that are characterized by risk management challenges and potentials for loss of primary containment (LOPC).

In these industries, contractors and service providers fulfill roles in meeting the business need for specialized services and skills when required by the organization, thereby relieving the organization of bearing the carrying cost of these specialized services and SMEs throughout the year. The authors recognize the following additional reasons for contracted services and materials providers:

- Cost management: Cost applicable to the organization as and when required. Competitive bids also help organizations to derive the best cost for the services and materials required.
- Contractors provide a pool of competent and capable human resources to select from when required.
- Contractors provide best practices, knowledge, and expertise from working with different clients often to the benefit of the organization.
- Contractors and services providers are easily changed when and if the organization is not happy with the services and materials provided.

For all these reasons, contractors are assuming an increasingly important role in business and should be managed to provide the greatest value for the organization.

13.4 Developing the scope of work

Scope of work (SOW) development requires stakeholder involvement in identifying the specific requirements to be communicated to the contractor or service provider. Stakeholders typically include the following:

- Engineering organization: Provides technical and engineering requirements for the specific SOW. Usually defined in engineering standards and specifications.
- Quality management organizations: Specifies the quality requirement for the specific SOW.
- Health, safety, environment, and security: Provides owner specific requirements defined in standards that should be met or exceeded.

- Procurement/supply chain management: Coordinates the process to ensure all documents required to ensure adequate information is communicated to the contractor to formulate bids.

Fig. 13.3 provides an overview of deliverables of both the owner and contractor in SOW development.

13.5 Internal/external prequalification

Owner prequalified contractor database is developed by screening out contractors who do not meet the owner's threshold requirements from its HSE and quality perspectives. A prequalified database of contractors may be maintained internal to the organization or can be provided by external prequalification service providers. Once the SOW is developed and requirements from the contractor clearly defined, the organization (owner) should invite only prequalified contractors to bid on the SOW.

Nonprequalified contractors are invited to bid only in exceptional circumstances to enhance competitive bidding when the group of prequalified service providers is limiting or when specialized services are provided only by a single service provider. A waiver process should be followed to ensure this process does not become normalized. Senior leadership approval of the waiver should apply to discourage the use of the waiver process.

13.6 Site visit verification

For major contracts, the organization may seek to conduct field visits to the sites of potential prequalified contractors before inviting them to bid. In this way, the organization gets the opportunity to better understand the HSE performance, work processes, resources capabilities, and the ability to deliver the SOW requirements. Of particular importance is the need to speak to workers who can provide realistic feedback on the contractor's or service provider's behaviors under normal and unprepared modes of operation.

13.7 References checks

Performing reference checks on prequalified potential bidders is an important step in the process to ensure the organization invited only suitable and competent contractors or service providers to bid on the

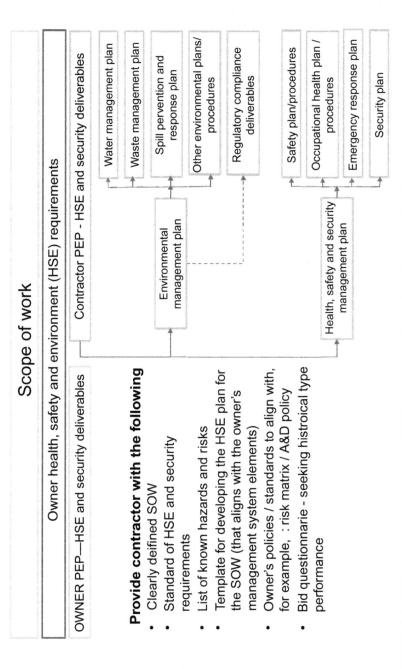

FIGURE 13.3 HSE and security scope of work (SOW) development. Source: *Safety Erudite Inc. (2019).*

SOW. Getting the feedback from prior clients of the contractor or service provider is an important step in screening out potential invitees to bid on the SOW. Reference checks should center around the following:

- HSE performance,
- quality performance and rework requirements, and
- schedule and cost management.

With effective HSE, quality and schedule management, budgeting, and cost management is generally a lot easier to manage. The organization should seek information from personnel directly involved in supervising and managing work performed by the contractor or service provider to obtain the most accurate performance information.

13.8 Selection of contractors: criteria and weighting and contractor selection criteria

Selecting the right contractor is the outcome of the combined effort of a series of key stakeholder groups that generally include the following:

- technical disciplines;
- quality management;
- HSE; and
- other specialized stakeholder groups based on the SOW definition.

Each stakeholder group is required to establish evaluation criteria for assessing each contractor or service provider. Sample criteria are provided in Table 13.1.

TABLE 13.1 Evaluation criteria—sample.

Technical (e.g., engineering)	Quality	HSE
• Design capabilities • Equipment and machinery • Trained and competent technical staff—resumes • Engineering support teams and interface organization • Engineering standards • Construction manuals • Project manager • Other	• Quality assurance manual • Quality procedures • Trained and competent staff—resumes • Other	• HSE plan for SOW • Preliminary hazard assessment • Procedures for SOW • Organization resumes • Responsibilities, resources, HSE policy standards, and documentation • Risk management • Implementation and performance monitoring • Management review • References and site assessment • Other

TABLE 13.2   Predetermined weighted score—sample.

Stakeholder group	Allocated weight %
Technical	40
Quality management	20
HSE	25
Other	15
Total	*100*

13.9 Stakeholder weighting assignments

Each stakeholder group is allocated a predetermined weighted score (as shown in Table 13.2) based on the relative importance of the stakeholder's group value proposition in the SOW. Further detailed stakeholder criteria weighting may be included as a detailed evaluation is undertaken based on predetermined selection criteria for each stakeholder group.

13.10 Health, safety, and environment evaluation

During the bid exercise, the contractor or service provider is asked to provide specific information based on which all contractors/service providers are to be assessed and evaluated. Fig. 13.4 shows the detailed HSE evaluation is performed based on the owner's HSE evaluation criteria.

Each stakeholder group may undertake a similar evaluation of contractor documentation and information supporting the bid.

13.11 Veto rights

>Occasionally, for high-risk SOW, veto rights may be applied over all other stakeholder selection criteria. After the HSE evaluation is completed, regardless of the scoring of other stakeholders, the contractor or service provider with the best HSE evaluation is selected over all other contractors. In such situations, this does not mean the evaluation of other stakeholders is not performed. Rather, the evaluation is completed by all stakeholder groups, but HSE scoring overrides all other stakeholder scorings.

Contractor HSE Evaluation - Weight 25%
Date:

Insert Evaluation Score Here Score 1-5 where 1= Poor and 5 = Excellent

Evaluation Criteria	Weighting	Contractor 1		Contractor 2		Contractor 3		Contractor 4	
		Score	Weighted Score	Score	Weighted Score	Score	Weighted Score	Score	Weighted Score
EHS Plan for SOW	20		0		0		0		0
Preliminary Hazard Assessment	5		0		0		0		0
Copies of Procedures for SOW	10		0		0		0		0
Org Chart and Resumes	2.5		0		0		0		0
Leadership and commitment	5		0		0		0		0
EHS policy and strategic objectives	2.5		0		0		0		0
Organisation, responsibilities, resources, standards and documentation	10		0		0		0		0
Risk management	10		0		0		0		0
Planning and procedures	5		0		0		0		0
Implementation and performance monitoring	10		0		0		0		0
EHS auditing	5		0		0		0		0
Management review	5		0		0		0		0
References & site assessment	5		0		0		0		0
Other	5		0		0		0		0
Total	100		0		0		0		0

Contractor with the highest weighted score has the best HSE Potential. Selection order shall be from highest weighted score to the lowest.

FIGURE 13.4 Detailed HSE evaluation performed by the organization. Source: *Safety Erudite Inc. (2019)*.

13.12 Commercial assessment

Assessment of the commercial aspects of contractors and service providers is generally retained by the supply chain management or procurement organization that administers the process. Generally, the financial aspects of the bid are not provided to the stakeholder until a contractor or service provider is selected based on a full assessment of all stakeholders. In most instances, however, cost pressures within organizations are often lead to the selection of contractors and service providers with the lowest bid price.

When this occurs, and particularly as it relates to HSE concerns, if the organization's HSE requirements are not met or HSE performance is likely to be compromised, negotiations with the organization may be required to augment HSE requirements with internal resources. What this means is that additional HSE support can be brought in to ensure the work is performed safely. In such instances, the cost savings by selecting a contractor or service provider with less than acceptable HSE capabilities should significantly outweigh the additional cost in procuring internal or external augmentation HSE support.

Such costs should be evaluated and added to the overall project cost.

13.13 Risk ranking of contractors

Understanding the risk presented by the contractor is of critical importance to the organization. A process to determine the risk presented by the contractor or service provider is required. Fig. 13.5 provides a simplified process for categorizing contractors based on the risk gaps in the HSE plans provided by the contractor for the contracted SOW. With each category of contractors or service providers, the organization's approach to and relationships with the contractor and/or service provider is adjusted to minimize risk exposures.

By using this matrix, the organization's approach to managing the contractor or service provider can range from directing and guiding the contractor when the risk ranking of the contractor is unacceptable through support only where the risk rank is low and assessed as local attention item.

13.14 Contract execution

Once the contractor has been selected, and the contract awarded the next steps involves the execution of the contract. Consistent with Fig. 13.2, there are several key activities involved in the process as follows:

- contract/project kick-off,
- contractor premobilization,
- the contractor provides an SOW-specific HSE plan,
- holding an HSE Alignment Workshop,
- site-specific training implemented,
- contractor mobilized, and
- contractor performance management.

Understanding the requirements around the above activities helps the organization to improve the overall contractor management and, in particular HSE management processes.

13.15 Contract/project kick-off

Once a contract is awarded and before fieldwork starts, there is a contract (project) kick-off meeting that is generally held between the organization's evaluation and procurement stakeholders and the contractor's leadership organization that are involved in developing the bid and contract agreements. This is the first opportunity for the organization to reinforce its requirements from the contractor regarding its priorities of the following order:

1. HSE requirements and expectations;
2. quality and engineering requirements;

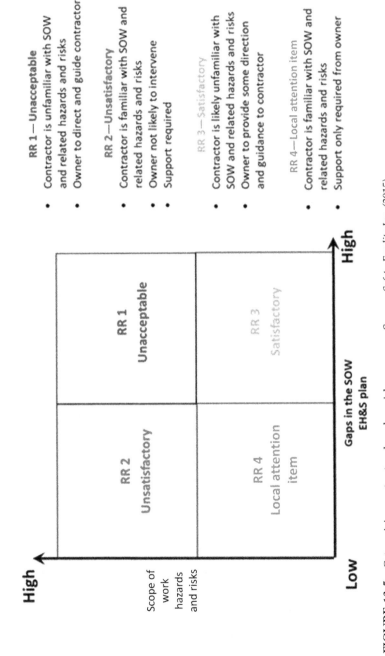

RR 1 – Unacceptable
- Contractor is unfamiliar with SOW and related hazards and risks
- Owner to direct and guide contractor

RR 2—Unsatisfactory
- Contractor is familiar with SOW and related hazards and risks
- Owner not likely to intervene
- Support required

RR 3—Satisfactory
- Contractor is likely unfamiliar with SOW and related hazards and risks
- Owner to provide some direction and guidance to contractor

RR 4—Local attention item
- Contractor is familiar with SOW and related hazards and risks
- Support only required from owner

High

	RR 2 Unsatisfactory	RR 1 Unacceptable
	RR 4 Local attention item	RR 3 Satisfactory

Scope of work hazards and risks

Gaps in the SOW EH&S plan

Low

High

FIGURE 13.5 Categorizing contractors based on risk exposures. Source: *Safety Erudite Inc.* (2015).

3. schedule deliverables; and

4. cost management.

When steps 1–3 are properly managed, cost management becomes a much easier exercise that generally leads to within or on-budget delivery of the project.

13.16 Contractor premobilization

The contractor premobilization meeting is an opportunity for the owner to verify the contractor's equipment and machinery, documented processes, personnel, and overall readiness to undertake the SOW. The premobilization process requires all owner stakeholders involved in the SOW to verify and document the contractor's readiness to be moved to the worksite. This activity occurs before the contractor is allowed to move to the worksite. Organizations should develop standardized premobilization checklists for each stakeholder group to be used before contractors are mobilized. This step needs to be built into the project's overall schedule so that sufficient time is given to address this without exerting undue pressure on the project schedule.

The contractor must address gaps in the premobilization requirements before they are allowed to move to the organization's worksites. A follow-up premobilization meeting may be required to ensure all gaps are properly addressed before mobilization. Premobilization should include, at a minimum, verification of the following:

- Training and competency verification of workers involved in delivering the SOW. Where applicable organizational augmentation may be required.
- Equipment and machinery inspections and licenses review.
- A review of the SOW-specific HSE plan that addressed SOW known hazards and risks. The plan should also address environmental hazards and risks associated with the work.
- Organizational chart of the contracting workforce and the owner/ contractor interface points.

13.17 HSE Alignment Workshops

Why is an HSE Alignment Workshop required? In most instances, when contracts are approved, senior leaders of both the owner and the contractor are involved in the process. Also, the HSE plan for the SOW is often a generic version that has been sitting on the shelf and dusted off for the bid process with some minor modifications. Furthermore, the project management team is often not very involved until the contractor is mobilized.

An HSE Alignment Workshop is required to ensure the contractor's project management workforce is fully aware of the owner's HSE requirements and expectations when the contractor is on the owner's worksites. Led by the owner and with the full participation of the contractor's leadership and HSE organization involved in the project, the owner details its HSE expectations and requirements. This is an ideal opportunity to reinforce safety first, and a value such as no work is so important that workers cannot take the time to do it safely. Moreover, the work priorities are always safety, schedule, and cost, and should be prioritized in that order.

13.18 Site-specific training implemented

Prior to mobilizing the contractor or service provider, all site-specific training should be completed by the owner. Site-specific training by the owner may include the following:

- unique hazards at the worksite;
- site life-saving rules and principles;
- site policies and expectations; and
- any site-specific technical training that may be required for the SOW.

Site-specific training is generally led by the owner's personal and retained onsite for future applications and similar or repeat of the SOW.

13.19 Contractor mobilized

Bearing in mind that the two times most incidents tend to occur with contractors are during the mobilization and setline in period and the demobilization period, careful owner attention is required during this period to minimize incidents. Vigilance by both the contractor and the owner is required to prevent incidents. Both the contractor and owner are required to provide adequate supervision and control to ensure work is performed as per procedures and with the right levels of hazard assessments and risk mitigation.

Reasons provided for an increased number of incidents during this period include the following:

- worker excitement with the new job,
- changes associated with a new worksite,

- need to learn quickly and often relearn a new way of doing the same thing, and
- inadequate supervision.

Careful attention to these concerns helps to start the contractor off on the new SOW with greater confidence and an opportunity to better succeed in the delivery of the SOW on-budget and on-schedule with excellence in HSE performance.

13.20 Contractor performance management

Once a contractor or service provider is successfully mobilized, the performance of the contractor must be managed to ensure safe, on-schedule and on-budget delivery of the SOW. Contractors must be assessed on a continual basis, independent of prior performance. Malhotra [2] advised, regardless of if a contractor has "two, 20 or 200 workers on site, the same assessment process must be consistently applied, albeit with varying levels of scrutiny relative to risk" (p. 29). Contractor performance management is the joint effort of the owner and the contractor. With preestablished key performance indicators (KPIs) that are communicated to the contractor and jointly agreed with, the owner and contractor jointly measure performance against these KPIs at an agreed frequency. Managing contractor performance is an ongoing process that requires trust, communication and ongoing continuous improvements in the execution of work.

Contractor performance management is enhanced when the owner performs the following:

- HSE audit: 30 days' postmobilization;
- leadership visibility;
- listening moments;
- contractor audits;
- corrective action management; and
- relationship management.

When done effectively, contractor performance is enhanced, leading to positive outcomes from the contracting relationship.

13.21 HSE audit: 30 days' postmobilization

Why is a contractor HSE audit required 30 days' postmobilization? For many organizations, this may appear to be too early in the process. However, the benefits of this audit are tremendous in the early identification of gaps in the HSE performance and management system of the contractor. If this exercise is done much later in the process, there is every likelihood of a strained relationship outcome between the contractor and the owner when the contractor sees any change as a deviation for the SOW requirements and expectations. The longer the period between mobilization and identification of gaps and implement the action of corrective actions, the more difficult the process.

13.22 Leadership visibility

Perhaps the most important thing the owner can do to ensure the best contractor performance is by being visible. Leadership visibility from both the contractor and owner organizations is extremely valuable in keeping HSE performance high and schedule, quality, and cost on target. Petronas [3] pointed to the company's safety initiatives for that year focused on leadership, "visibility and communication to influence HSE behavioral changes at all working levels, including contractors" (p. 4). Leadership visibility includes the effective engagement of workers in the field. This engagement should focus on the worker's understanding of the work being undertaken, hazards associated with the work and the leader's demonstrated genuine care and empathy for the worker regardless of whether an employee or contractor is engaged. Interserve [4] advised that proactive site visits, safety inspections by the executive board, directors, management teams, and safety advisers led to >131,500 unsafe conditions being identified and corrected, thereby preventing potential incidents.

Joint contractor and owner leadership visibility are the most effective means of improving contractor performance. During such visits, the same messages are communicated to workers by both the contractor and the owner's leaders. Not a common practice, organizations are increasingly encouraging its leaders to perform joint site visits as a means for improving contractor HSE performance. Joint leadership visibility at higher levels of the organization is more effective in improving contractor HSE performance.

13.23 Listening moments

What are listening moments? Listening moments are opportunities provided by the owner to its contractors for them to provide information, feedback and potential improvement opportunities without documentation of discussions. In most instances, when contractors are averse to talking when the information provided by the contractor is documented. This is particularly so when multiple contractors are working on the same site. For listening moments to be effective, leaders should gather all contractors in a common area and ask questions around themes. Listening moments can be done around lunchtime, where the owner provides lunch and creates an informal meeting environment. Contractors are encouraged to speak freely without fear of repercussions for identifying and reporting concerns. In these events, there is no documentation of the meeting and discussion.

The owner encourages contractors to talk by asking probing questions consistent with the following themes:

- What are known areas of vulnerabilities that can lead to incidents?
- What can be done better to improve HSE and contract performance?

When information is provided, the owner must act to address the concerns and vulnerabilities identified. By doing so, credibility is developed, and contractors become more motivated to bring forward concerns that can lead to incidents if left unaddressed. Listening moments is an effective means for proactively preventing incidents in the workplace.

13.24 Contractor audits

The contractor should be audited relative to the SOW commitments and deliverables. Owners may audit the contractor's processes to ensure commitments are maintained so that safe, reliable and efficient work is performed. Collaborative audits are encouraged whereby contractors actively participate in the audit and are encouraged to surface problems and issues before incidents occur. Invasive audits, on the other hand, encourage contractors to hide problems and concerns that may eventually lead to incidents and adverse performance impacts. Fig. 13.6 demonstrates the impact of collaborative auditing of contractors and HSE performance improvements.

The impact of collaborative field audits in large spend oil and gas contracting organizations over a 2-year period led to ~100% improvement in the contractor recordable injury frequency.

Audits of contracting organizations should adopt a disciplined approach to focusing on the following:

- findings relative to requirements;
- gaps in the process: generally within the categories of procedures, accountability, competence, and assurance; and
- corrective actions management.

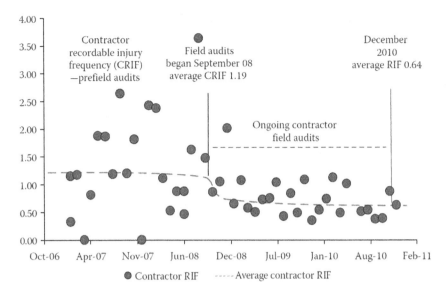

FIGURE 13.6 Collaborative auditing of contractors: HSE performance improvements [5].

Audits should be risk-based and performed at the right frequency to avoid audit fatigue and overtaxing the business in performing audits to check the box for having completed an assigned activity.

13.25 Corrective action management

Corrective actions should be prioritized, assigned, resourced, and stewarded to completion. Corrective actions may result from the following sources:

- incidents investigations and root cause analyses;
- inspections;
- observation programs: Unsafe acts and behaviors, and unsafe conditions;
- audits; and
- risk management.

All corrective actions should be prioritized based on risks and exposures from the gap and assigned to a responsible individual with a defined period to close the gap. All high priority gaps must be addressed within the period defined (particularly where consequences can be high) to create and sustain trust in the process.

13.26 Contractor relationship management

Contractor relationship management depends on creating win−win situations and outcomes from the contracting process. Active engagement, involvement, fair treatment, and an environment of transparency are essential for creating trust and strong contractor management relationships. Managing relationships with contractors allows the development of symbiotic relationships between contractors and owners. These relationships are developed when leaders meet, discuss, treat each other with mutual respect and work together as a team.

Contractor relationship management is an important variable in creating a learning environment for organizations. Leif Jarle and Kåre [6] advised important factors for failure-based learning, both directly and through increased knowledge exchange is facilitated by contractor relationship management, among other variables such as leadership involvement, role clarity, and empowerment. When strong relationships exist, problems are surfaced and resolved before incidents, or cost escalations arise. Contractor relationship management requires work from both sides. When done properly, the benefits to both parties are tremendous.

13.27 Contract closeout

Many organizations forget to complete these stages of the contract life cycle. Closing out the contract in a timely manner and before the contractor, demobilizes provide opportunities for both parties to agree on the following performance measures:

- HSE and quality,
- cost and schedules,
- successes and failures, and
- best practices and key learnings.

Contract closeout is done using a structured process and should be done jointly with the contractor. Information gathered from the process should be used to close the loop in the prequalification database that will be useful in determining whether the contractor will be used again in future similar projects.

References

[1] K. Keay, Plugging into the GIG economy. Five reasons why contract work is alive and well in the lighting and electrical industry, Lighting Des. Appl. 48 (8) (2018) 24. Illuminating Eng. Soc. North. Am. Retrieved June 11, 2019, from ProQuest Database.
[2] S. Malhotra, Understanding the contractor management paradox, Acad. J. Prof. Saf. 64 (5) (2019) 27−30. Retrieved July 03, 2019 from EBSCOHost Database.

[3] Petronas Gas Berhad, Annual report 2006, Reportal Co. Rep. 2006 1st Quart. (2006). p. 1. 105. Retrieved July 03, 2019, from EBSCOHost Database.

[4] Interserve, Annual Report 2016. Reportal Company Reports. 2016 4th Quarter, Preceding p1-190, 2016. Retrieved July 03, 2019, from EBSCOHost Database.

[5] C. Lutchman, R. Maharaj, W. Ghanem, Safety Management. A Comprehensive Approach to Developing a Sustainable System, first ed., CRC Press, 2015, ISBN: 9780429107207, eBook.

[6] G. Leif Jarle, H. Kåre, Knowledge exchange and learning from failures in distributed environments: The role of contractor relationship management and work characteristics, Reliab. Eng. Syst. Saf. 133 (2015) 167–175. Retrieved July 03, 2019, from EBSCOHost Database.

Emergency response management and control

Ahmed Khalil Ebrahim

Bahrain Petroleum Company (Bapco), Kingdom Of Bahrain

14.1 Introduction

This chapter will provide a general overview of the emergency response plan (ERP) and system in addition to policies required to ensure efficient utilization of resources and effective management of emergencies. The chapter will describe, in general terms, basis of Incident Management System (IMS), incident classification and how scalable response can be initiated, how response organizations are structured, mechanism utilized to triggers activation of a response, incident management processes, linkage to business continuity, and other frameworks.

The chapter will also cover mutual aid aspect of the emergency, dealing with external stakeholders, media plans, drills and exercises in addition to recovery and lesson learned. ERP is a fundamental element and an integral part of any process safety management system for any organization.

14.2 Why have emergency response plan

No emergency can be managed without having a detailed process that explain how emergencies and crisis will be managed taking into consideration the type of risk and the impact of an emergency on the establishment and concerned stakeholders. As a minimum the ERP will

Process Safety Management and Human Factors.
DOI: https://doi.org/10.1016/B978-0-12-818109-6.00014-9

describe the expectations, scope and content of IMS or as it referred to in this section as the ERP. The IMS will:

- Provide guidance to Emergency Response Teams, Incident Management Team, Crisis Management Team and those supporting them for the response to and control of an emergency, incident, or crisis associated with the organization facilities or personnel.
- Explain the response organizational structure, its functions, and the roles and responsibilities of key personnel within it.
- Define response team notification and activation procedures.
- Describe the planning processes for, and implementation of, emergency, incident, and crisis management.
- Depict the linkage of crisis and continuity plans and the functions responsible for invoking and implementing the plans.

14.3 Scope of the Incident Management System or emergency response plan

While the emphasis is on incident management at Level 2 (Tactical Response), it is important to recognize that the scope of the ERP is much wider. The Command and Control (Chain of Command) consist of three Scalable Response Organization Level Structure which is designed to ensure a standardized approach to Emergency Response and Crisis Management. The three levels are

Level 1: On-Scene Commander and Emergency Response Teams "Operational Response";

Level 2: Incident Management Team (IMT) "Tactical Response"; and

Level 3: Crisis Management Team (CMT) "Strategic Response."

This scalable response structure has the capability of expanding, contracting or substituting expertise within the response organization to match the complexities and demands of single, multiple or complex incidents. The ERP normally would acknowledge the broader concept of crisis management and the crisis management processes that lead to successful outcomes in relation to internal or external situations or events impacting the business. The ERP would also normally outline the management of facility level response addressing all areas, assets and the procedures to respond to incidents.

The effectiveness of any ERP will depend heavily on how it is articulated, continuously practiced, updated but also it relays on having the right resources, at the right place with the right experience.

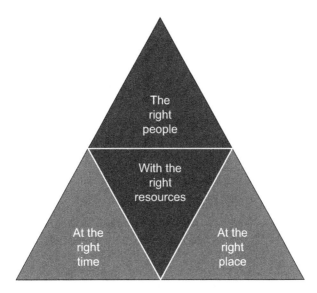

Most organizations and government entities around the world have opted to adopt one form of a structured way of managing the various types of emergencies and in most cases, they utilize an integrated structure which is referred to Incident Command System (ICS), which is widely used within the process industry especially in the oil and gas industry. ICS is a standardized all-hazards incident management concept which incorporates a set of proven organizational and management principles, terminologies, processes, and procedures.

ICS was developed in the United States for the management of incidents, which required interagency responses and coordination. The system was carefully reviewed by the US Federal Emergency Management Agency (FEMA), in recent years, following the Gulf of Mexico incident and was refined accordingly incorporating all lessoned learned and making sure it meets the intended purpose.

The system allows for a wide range of response organizations, contractors, and resources to be called to respond to incidents with varying missions and procedures. The coordination of and collaboration between these organizations is critical to an effective response operation. These groups and individuals must be able to work together at short notice and may have little or no prior experience of collaborating with each other to manage stressful, dangerous and evolving problems in what may be a hazardous working environment.

Responders are required to cultivate a working trust with one another, have clear roles, responsibilities, and authorities, and ensure that sufficient on-scene resources are available always. Responders face many other potential challenges in responding effectively to major incidents. Factors such as weather, site access, resource constraints, poor coordination, or poor communications can delay response times or hinder the efficiency of the response. That is why ICS is an essential tool for overcoming many of these challenges; it provides clarity in command and control, improves resource coordination and communications, and facilitates the cooperation and integration of the various parts of the response organization.

The strength of ICS is its ability to be scalable to suit the type and size of the response and it deploys a systematic method for coordinating and controlling a variety of activities, resources, and response organizations from a command center. This is to ensure that all personnel work to a common system and objective and that they are familiar with the various roles, responsibilities, and processes of the system.

Other benefits from utilizing ICS system include but are not limited to:

- Actions of the teams involved in the response are documented, this allows for traceability of decisions and ensures accountability of resource deployment, workloads, and demobilization of personnel and equipment.
- Documentation allows for more accurate assessment of the "as is" situation, thus allowing recovery operations or the move from emergency and crisis management into business recovery or "project" to be expedient and seamless.
- Full records are kept of all communication including incurred costs and the cost of recovery under Insurances.

It is essential for the successful implementation of ICS during emergencies to ensure effective coordination and cooperation between the on-scene commanders, the incident management team (IMT), incident commander (IC), the CMT and external or governmental representatives. It must be appreciated that the IMT IC maintains the ultimate authority over the IMT response and resources for incidents within their sphere of responsibility; however, the severity of the incident may extend to the outside boundaries which will dictate the overall level of involvement of the IC.

If the incident escalates to level 3 (Tier 3) and if it has impact on the company or the surrounding communities, then it is essential to establish a crisis management team (CMT) led by a crisis manager (CM) who must be senior enough to make high level decisions and has the authority to involve external agencies if required. This appointed individual would have a much wider role as he would have to consider the holistic impact on the company and the surrounding communities and even on the country. It is important for him to ensure that all response actions take into consideration external agencies plans and command structure and the manner they are integrated with the overall crisis management system.

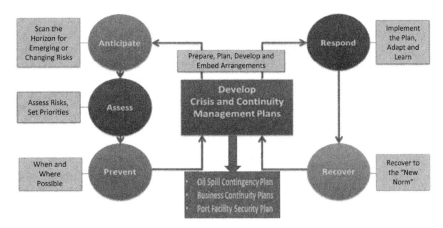

FIGURE 14.1   Integrated "all-hazards" approach to incident management.

14.4 Integrated "all-hazards" approach

If organizations seek to create in place effective and efficient integrated (all hazard) response plan that takes into considerations, planning, activation, response, and recovery than it must be based on business resilience management system framework (BRMS). The BRMS provides for an integrated "all-hazards" approach to anticipating, assessing and preventing hazards, and developing and implementing response plans to effect recovery. The system emphasizes that all response plans must be firmly rooted within the context of the existing response organization and its processes to prevent each appearing to be single entities (Fig. 14.1).

A key component of the "all-hazards" approach to incident management is to provide an effective organization with the necessary functionalities and connectivity to deal with any potential emergency that endangers people, the environment, company assets, or business reputation.

14.5 Organizational principles of all-hazards approach

The principles of the hazard assessment approach can be summarized in the following:

- Use of a single, integrated organization to manage the response.
- Organization by function, that is, command, operations, planning, logistics, and finance.

- Establishment of clear, hierarchical reporting relationships.
- Maintaining operational readiness of a modular and scalable organization and ensuring that it is appropriately sized to achieve the response objectives.
- Philosophy is based on assessment of the risk, respond, manage, adjust (up-scale/down size/substitute expertise) and stand-down.

14.6 Emergency response priorities

In an Emergency Response and Management plan, the following priorities must be adhered to at all stages of managing the emergency; and in this order:

- *People*: Protect the health and safety of responders and the public.
- *Environment*: Protect and mitigate impacts to the environment.
- *Assets*: Protect the organization assets and public infrastructure from any further impact.
- *Reputation*: Protect the organization reputation/business by ensuring response activities are not detrimental to business continuity and are conducted ethically and transparently.

14.7 Emergency management principles

Management principles provide the IC and command staff with guidelines to coordinate the efforts of the organization so that response objectives and priorities are accomplished through the efficient and effective use of the available resources. Management principles include operational planning, staffing, leading, directing, and controlling the organization. In general terms, it is based on the following principles:

- ensuring an objectives-driven response;
- formulation of incident action plans;
- use of common, consistent, and understood terminology;
- maintaining a manageable span of control; and
- effective coordination of equipment, personnel resources, and communications.

14.8 Objectives-driven response

An effective and successful response requires clear objectives; these objectives are established at all levels. The CMT will focus on strategic objectives while the IMT will normally concentrate on tactical objectives as they drive the development of response strategies which are developed and implemented through tactical decisions and actions on the ground (work assignments developed against the tactics and assigned to individuals or teams). The objectives, strategies and tactics evolve as the response progresses. Objectives at all levels of response should be based on the SMART principle: *Specific, Measurable, Action-oriented, Realistic,* and *Timely.*

14.9 Incident action plans

Another important part of ERPs especially when ICS is adopted is the creation of incident action plans (IAPs) which is designed to provide control of response activities for a specified period. IAPs ensure that all parts of the response organization work in coordination toward common goal and to achieve the objectives set by the commander. IAP describes the overall objectives and strategies for managing the response as well as response tactics for a set length of time, known as the operational period. An IAP includes the identification of operational resources, and provides a documented record of work assignments, priorities, safety, and environmental considerations.

14.10 Common and consistent terminology

One of the strengths of ICS is that it employs common terminologies which are used to prevent misunderstandings when responding to an incident. Common terms allow different parts of the response organization to work together effectively, and to communicate clearly with each other on essential issues:

- The terminology used for each organizational function is standard.
- Major resources (personnel and equipment) are given common names.
- Incident facilities are named according to common terminology.
- Position titles: All responders within identified "teams" are referred to by standardized titles, such as commander, officer, chief, director, supervisor, or leader.

14.11 Manageable span of control

Another crucial part of any effective ERP especially ICS is the manner resources are managed, this is referred to as span of control, which is the number of individuals or resources that can effectively be managed by a supervisor during an incident. A recommended span of control ranges from three to seven individuals, with five representing the optimal level. There may be exceptions to this range, for example, in cases of lower-risk assignments, or assignments requiring minimal direct supervision.

14.12 Organizational response structure

14.12.1 Incident classification and aligned scalable response

The following is an example of incident classification and the scalable response against each classification (see Table 14.1).

See Fig. 14.2 for tiered emergency response structure.

14.13 Scalable response

Table 14.1 demonstrates the tired response and is aligned with scalable response and Incident Classification Matrix, which allows the response organization to be structured in a way that is appropriate for the size and complexity of the incident.

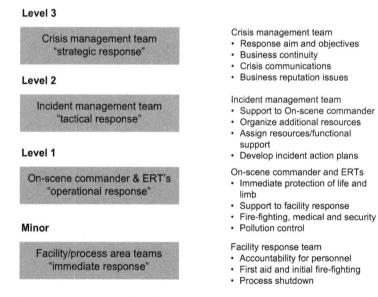

Level 3

Crisis management team
"strategic response"

Level 2

Incident management team
"tactical response"

Level 1

On-scene commander & ERT's
"operational response"

Minor

Facility/process area teams
"immediate response"

Crisis management team
• Response aim and objectives
• Business continuity
• Crisis communications
• Business reputation issues

Incident management team
• Support to On-scene commander
• Organize additional resources
• Assign resources/functional support
• Develop incident action plans

On-scene commander and ERTs
• Immediate protection of life and limb
• Support to facility response
• Fire-fighting, medical and security
• Pollution control

Facility response team
• Accountability for personnel
• First aid and initial fire-fighting
• Process shutdown

FIGURE 14.2   The tiered emergency response structure.

TABLE 14.1 Incident classification matrix.

Level	Incident description	How incident is handled
MINOR	*Managed by: owner department/facility/site response team and equipment*	
	An incident which can be dealt with by the immediate facility/location personnel and resources.	This incident should be managed as per the owner facility/ site specific response plans.
	The incident does not have any effect outside the facility/location with <u>no</u> external involvement.	
	There is <u>no</u> risk to life, to the environment, assets or reputation.	
LEVEL 1	*Managed by: on-scene commander Emergency Response Teams (ERTs)*	
	An incident which can be dealt with by the facility or location personnel and resources supported by the On-Scene Commander and Emergency Response Teams	The incident should be managed by the ERTs.
	The incident does not have any effect outside the facility/location with <u>limited</u> external involvement.	
	There is minimal <u>risk</u> to life, to the environment, assets or reputation.	
LEVEL 2	*Managed by: Incident Management Team (IMT)*	
	An incident which may be dealt with locally but requires support from Level 2 (IMT) organization. It may involve external resources and agencies.	Requires mobilization of the Incident Management Team (IMT) to a level appropriate to the scope and scale of the incident and its potential to escalate (at a minimum the Core IMT staff will activate).
	The incident may be onsite, have some effect outside the site or could be offsite.	Requires only notification or minimal direction/input from the Crises Management Team (CMT).
	There is a potential risk to life, to the environment, assets or reputation.	
LEVEL 3	*Managed by: Crisis Management Team (CMT)*	
	An incident, which requires the use of wide-ranging resources and/or government and external agencies:	Requires full mobilization of the IMT and CMT.
	The incident will have technical, media and external affairs which require immediate assistance from the IMT and from the CMT.	Potential for Coordinated or Unified Command with Government/ External Agencies.

(Continued)

TABLE 14.1   (Continued)

Level	Incident description	How incident is handled
	There will be at least one of the following: • Death and/or serious injuries; • Potential for environmental damage; • Substantial damage to property; • Damage to the Company reputation.	

14.13.1 Scalable response organization

14.13.1.1 Duty management

Most Organizations have in place an "Emergency Response Duty Call Out System," which is designed to notify all those who are on call out and to ensure that the correct classification of any event is made quickly and accurately and that the correct response resources are deployed to manage the event.

The response teams build from the bottom up, with responsibility and performance placed initially with the duty shift supervisor who acts as the initial on-scene commander (OSC). The OSC has full autonomy of operation and reports to the designated Senior Manager who would normally be the IMT IC if and when the IMT is activated, who is responsible for assessing and managing any consequences or potential impacts of the incident as well as providing support to the OSC through the IMT organization.

14.13.1.2 Facility incident response

Most organizations or facilities have a response team which is organized and trained to undertake Immediate Response to any incident.

14.13.1.3 Scalable response structure

Using the scalable structure as a basis, the ERP/ICS organization can expand and contract whilst ensuring that the critical incident management functions are addressed and that an appropriate level of response is provided to deal with specific incident types and the ever-changing requirements of ongoing incident management.

14.13.1.4 *Example of notification and activation of emergency response plan*

The diagram below outline process for the notification and activation of a typical company crisis and emergency response organization (Fig. 14.3).

14.14 Response triggers

See Table 14.2 for response triggers for a refinery.

14.15 Managing incident response through the utilization of the planning cycle

Regardless of the size and type of the incident, the response process begins with incident detection, notification, activation of the ERP and response teams and the establishment of incident command organization.

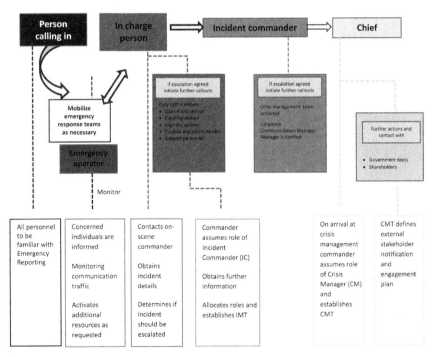

FIGURE 14.3 Notification and activation of crisis and emergency response organization.

TABLE 14.2 Response triggers for a refinery.

Safety and health	Regulatory and environment	Asset integrity	Business continuity	Business reputation	Community
Incidents impacting on people	Regulatory impact	Damage to processes or facilities	Inability to undertake normal business	Concern over company activities	Impacts on the community
Major Asset Fire Explosion	Offshore spill	Collision/loss of stability/structural failure at:	Interruption to export feed stock and supply of fuel	Unplanned media attention on Company Activities	Public or third-party fatality
Significant gas release	Land spill with threat to waterway, water body, public or other receptors	Damage to processing facilities, including;	Maintenance of Strategic Reserves for the Local Market	Involvement of NGO's	Loss of supply to Service Stations
Mass Casualty Incident	Release to air or sea with impact to personnel, public, environment	Supply of Cooling water	Refinery Lab Operations	HR Crisis*	Potential or partial community evacuation or major road closure
Fatality of Worker(s) through Industrial Accident	Regulatory Authority response Mobilized for the incident	Marine and Marketing Facilities	Road distribution of products	Strike* (*applicable also to Business Continuity)	Potential adverse health effect from impaired air quality or drinking water from any Company process or supply
Maritime Security threat	Probable impact to sensitive species, sensitive land				
Security threat requiring Facility Evacuation	Licenses to operate suspended	Distribution Depot	HR Crisis*	Significant impact on market share	Impact or potential impact to marine/land use or access to public places
pidemic/Pandemic		Loss of equipment critical to continued operations—Fire, Medical and Communications	Strike* (*applicable also to Business reputation) Provision of IT Services / Note: for more info please refer to BCP's in DMS		

IMT 'Initial Response'

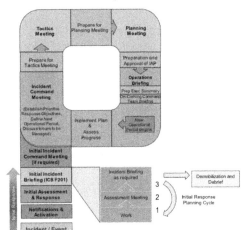

IMT 'Proactive Response'

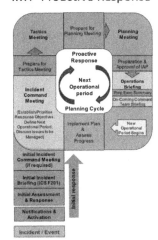

FIGURE 14.4   "Initial response" and "proactive response": planning "P" model.

If the incident develops from one level to another, the response and the IMT organizational structure and cyclical planning processes to bring the incident under control is established and implemented.

For small incidents, the initial response of the IMT and the planning cycle may be relatively simple. But for complex and protracted incidents, the IMT organization will be more structured and the planning cycle is more defined. The incident management process becomes proactive with comprehensive written IAPs, which include tactics and resource assignments to accomplish the response objectives and strategies established by the IC.

Most organizations would use the proactive response planning cycles, which is based on the ICS *Planning "P" Model* as outlined in Fig. 14.4.

14.16 Business continuity

14.16.1 Business continuity management (recovery stage)

Business continuity management (BCM) deals with the preparation, testing, and upkeep of business continuity plans (BCPs) and the invoking of such plans against disruptive scenarios. BCM and associated BCPs should be an integral component of any organization emergency response and crisis management plan and should be driven by the company risk register and response readiness priorities. These priorities and risk register should at least dictate that:

- critical risks against credible disruption scenarios require BCPs;
- all business processes impacted by critical risks require BCPs; and
- all mission-critical business processes: those essential to maintain core business require BCPs.

14.16.2 Invoking business continuity plans

14.16.2.1 Business continuity plans for foreseen or assessed disruptive scenarios

From incident management perspective the crises management commander has the authority to invoke BCPs against significant business disruption. BCPs can function as a singular entity (supported by the business resilience management team) and work with the affected processes or functions and the associated business recovery team or teams to reinstate the process or function to "normal" operations.

14.16.2.2 Business continuity plans for unforeseen or "nonassessed situations"

The CMT approach to unforeseen or "nonassessed situations" should be to determine the strategic aim, define the objectives and select suitable and sufficient recovery strategies to support a recovery plan. Furthermore, their mandate is to provide the resources required for the development and implementation of that plan.

14.16.2.3 Invoking business continuity plans during ongoing incident management

There will be situations where an incident or event has happened and the IMT decided to activate the BCPs in coordination with the authorized authority (CMT) and it could either be an integral part of the ongoing incident response or at the point where the incident response is being closed-out. It is against the "as is" situation and from the feedback provided by the IMT commander to the CMT commander in conjunction with the business owner to decide when the right time to invoke BCPs or a series of BCPs. Here the authority may be delegated by the CMT to the IC and BCPs managed by the IMT. The BCP team would normally continue to operate until the organization is capable to resume normal business.

14.17 Plan linkage

In managing an "all-hazards" response the OSC and ERTs, the IMT, and the CMT must can access and implementing the various response plans available within the organization. Response teams may only have to concentrate on one plan or for major and more complex incidents, integrate several plans, sequentially or collectively, into their incident management processes. An outline of key plans (oil spill contingency plans, business continuity plans, security plan, international shipping and port security plan (ISPS)) linkage aligned to the overall business resilience management system and the organization response structure is outlined in Fig. 14.5.

14.18 Application of the Incident Management System in varying response frameworks

There are three models under which incidents may be managed. Their application will depend on the nature and size of the response. These are defined as

- single command,
- coordinated command, and
- unified command.

14.19 Single command

Most organization incidents are managed directly ("in house") using their response teams and supporting contractors. The response is conducted using the scalable response structure, the planning "P" model and senior management decision-making processes. In some circumstances, there may be appropriate oversight by government agencies.

14.20 Coordinated command

This model can be utilized where government agencies and the company are responding in parallel to an incident. Coordination of activities and alignment in decision-making is achieved through close liaison between command centers and is facilitated by competent persons authorized to represent their respective organizations. The coordinated command structure is identified later.

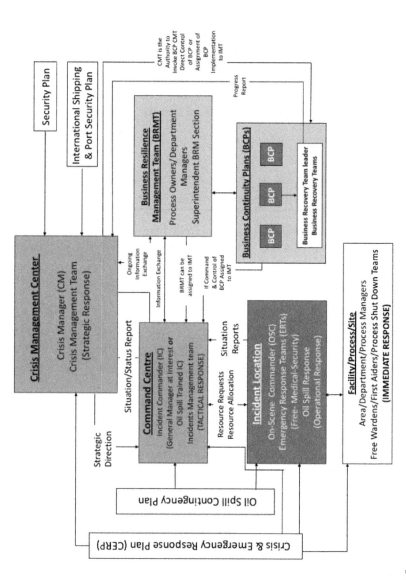

FIGURE 14.5 Business resilience management system: plan linkage.

14.21 Coordinated command structure

In respect of Coordinated Command during an offsite incident involving company facilities or where the company may be asked to take command or at least be heavily involved in the response. The response team and individuals in the team must ensure that they work with and support the authorities and other associated organizations involved in a response. All personnel involved in the incident should be kept appraised of who is actually in control of the incident at all times (Fig. 14.6).

14.21.1 Unified command

When an incident occurs, the initial views should be to have a highly focused single command response effort. Constructing such an effort can be difficult when multiple organizations exist with the authority to launch simultaneous, potentially divergent response operations. Unified command concept is designed to address these problems.

Unified command is usually implemented for larger type emergencies that have the potential to impact the country, natural resources/community activities or involve multiple jurisdictions. Unified command is a way of assuring that outside parties can provide mutual aid, and can participate in:

- determining overall response objectives and priorities;
- selection of strategies;

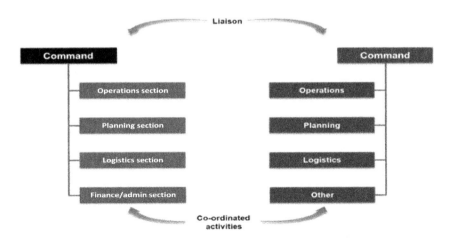

FIGURE 14.6 Coordinated command structure.

- maximizing available resources and the integration of response resources;
- joint planning for tactical activities such as major clean up and waste disposal;
- joint press releases; and
- resolution of conflicts of interest.

The makeup of the Unified Command structure depends on the type of incident, jurisdictions and resources involved; it does not, however, indicate that control of the incident facility and that it must be handed over to someone else. It is suggested that the priority should as much as possible be to retain the overall responsibility within the confines of the company operating site/business activity.

Mutual aid teams, for example, must come under the command of the company OSC or IC who is ultimately responsible for the incident site and all within its confines and boundaries. Where the incident encroaches on public areas such as main roads, for example, then it may be necessary to hand over control to local authorities.

The possibility of a unified command situation will only arise at a classification of Level 2 or 3 incident. The interaction between agencies involved in unified command is outlined later.

14.21.2 Unified command organizational structure

See Fig. 14.7 for unified command organizational structure.

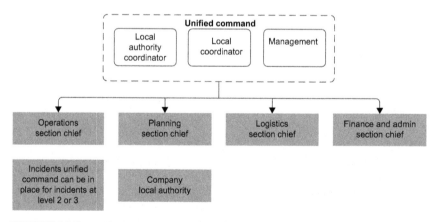

FIGURE 14.7   Unified command organizational structure.

14.22 Expectations and assumptions for the effective operation of the ERP or IMS

The following are expectations for the successful application of the IMS:

- Agencies and entities that support or coordinate with any organization must have good understanding of ICS, and have undertaken joint training and exercise with.
- Emergency management plans are based on risks that potentially impact the business. These plans are documented, accessible, clearly communicated, and aligned to the Company Business Resilience Management Systems and HSE Management System.
- ERPs, tools, and resources in place for each facility are kept up-to-date and are understood by those who may be affected and those who may have to respond.
- Assessments of potential emergency situations and their potential impacts are periodically conducted.
- Equipment, facilities, and personnel needed for emergency response are identified and available.
- Personnel are trained and understand emergency plans, their roles and responsibilities, and the use of the respective management tools and resources.
- Drills and exercises are conducted to assess and improve emergency response/crisis management capabilities, including liaison with and involvement of external organizations.
- Periodic updates of plans and training are used to incorporate lessons learned from previous incidents (internal and external) and exercises.

14.23 Human factors in emergency response planning

Human factors is an important part in effectively managing emergencies if not considered at the various stages such as the planning, alert, response, and even recovery. When designing emergency control rooms, it must take into consideration the operators function and the interaction he would have with the various tools. Ergonomic design, lighting and acoustic can help operator's carryout there assigned duties more efficiently and will be prone less faults. The way operators deal with callers requesting emergency services is another factor that could help gather as much information as possible which in turn can provide responders with adequate information that can help them deal with the situation more effectively in term of equipment and resources required but could also contribute to saving lives.

Understanding the issues related to humane factors with emergency responders is another important part. The use of response equipment, communication skills, dealing with spectators, data gathering, dealing with casualties and their relatives, incident investigation, and dealing with the public and the media are all important in managing the emergencies but also to the confidence and competency of the responders. These important traits are often ignored and not taken into consideration when training operators and responders and it is not mapped into their competency matrix and framework. It is only recently that some governmental organizations and companies that have built this into the training curriculum and uses when equipment are designed and when response procedures are being written. Experience shows that organizations who opted to build human factors in their emergency response strategy have become more successful in effectively dealing with emergencies and the recovery process. The above can only be achieved once human factors is fully understood and that an experienced individual is used to ensure that it is being incorporated in the various stages of emergency response including the design and training aspect. Those involved in such activities need to be trained and clearly made aware of it so that they understand the manner it is incorporated in their functions.

Human factors must also be considered when designing building and facilities especially when deciding on location of exits and escape routes. Smoke and fire behaviors and its control in buildings can hugely impact the severity and control of fires.

14.24 Concluding remarks

Countries and companies can have the best fire prevention systems in the world and they can provide the latest and most sophisticated equipment to protect their properties and facilities but would not be totally able to protect its assets unless they have a very well-articulated emergency preplans, comprehensive response procedure, adequate equipment designed or the types of risks and the trained competent responders. All of this is augmented with regular drills and exercises that are designed to deal with the various scenarios and risks in any given situations. They world have experienced many failures in effectively dealing with emergencies because either being over confidence of their capabilities and resources or because they were not adequately equipped, trained or even familiar with the various scenarios and risks in their own companies.

We have all failed to learn from other companies' major incidents and we did not carefully examine why those companies and countries have suffered major losses and why they did not deal effectively with those emergencies. These situations are golden opportunities for other entities to learn from the many mistakes and errors that they have experienced or made either in the way they managed the emergency or even the way the risk is assessed and managed.

Organizations must critically review the root of these incidents, look at the applicability of the recommendations and adopt them If possible. However, what would help organizations effectively deal with emergencies is having a robust response plan, comprehensive incident preplans, trained and competent responders, adequate and specialized equipment, regular drills and exercises in addition to regular reviews of the plans to ensure its adequacy and suitability to the various risks encountered by the company. The human element can be supported by training, rehearsing, and conducting regular drills and exercises as often as possible can help organizations ensure the effectiveness of their response plans and identify any weaknesses and areas of improvements.

ERPs as a dynamic process that needs be continuously improved and updated to reflect the current risk profile and to ensure that is in line with international best practices.

15

Human performance within process safety management compliance assurance

Janette Edmonds
The Keil Centre Limited, Edinburgh, United Kingdom

15.1 Introduction

Process safety management (PSM) and human factors ultimately share the same goal; that is to develop and maintain plant systems to prevent process safety incidents. Humans are often seen as the weak link in the system that cause adverse events, but there is a clear need to understand what is really behind the "human" weaknesses. It is true that human performance is less predictable than the engineering performance of a system, but the "human" weaknesses often occur as a result of how a system has been designed and engineered. Failing to properly integrate human factors within engineering activities and the PSM assurance process means that the unidentified or unacknowledged weaknesses will lie in wait, paving the way for the adverse event to occur.

There are numerous PSM techniques used from early system conception through to decommissioning, such as hazard identification and operability studies (HAZOPS); consequence evaluation; risk assessment; as low as reasonably practicable (ALARP) studies; prestart-up action response auditing; inspections; and auditing. The principles of PSM assurance apply throughout the life cycle of a plant or installation to ensure that the facilities can be managed safely and achieve acceptable levels of risk.

Process Safety Management and Human Factors.
DOI: https://doi.org/10.1016/B978-0-12-818109-6.00015-0
201

Likewise, there are numerous human factors techniques which are also used from concept through to decommissioning. These are applied to understand human behavior and to reduce risks from inadequate performance that can lead to process safety incidents. The application of human factors techniques enables assurance of the human elements within the system.

Applying PSM techniques is not the same as applying human factors techniques as they have different objectives. Typically, one is largely equipment focused and the other human focused. Even the inclusion of project activities such as HAZOP studies which have the "operational activity" implied in the title, are still insufficient to adequately assure the human element.

The PSM assurance process, although systematic and mechanistic, can at times fail to address the real underlying weaknesses that lead to significant loss of containment or otherwise serious process safety incidents. There is a need to integrate human factors within this process to enable the identification and management of vulnerabilities related to human failure.

This chapter provides an insight into why humans fail, the gaps within PSM assurance, and what can be done to assure human performance.

15.2 Sociotechnical systems and human failure

Although there is a need to understand "human" weaknesses, there is a need to recognize the strength that comes from having the human within the system.

There are many assumptions about human performance, and often a lack of forethought given to humans not performing as expected. Therefore, one of the roles of the human factors analyst during HAZOP studies and more generally is to challenge the perception of the human as always "performing as expected." Another role is to actively and systematically seek out human failure vulnerabilities.

In any sociotechnical system (i.e., where humans interact with technology), humans can fail in two ways, either because they intentionally or unintentionally do something wrong, referred to as a violation or error, respectively.

Violations generally occur because the person is motivated to do the wrong thing, for example, to save time or effort (routine violation); because they are unable to comply in the given situation (situational violation); or because exceptional circumstances force them to violate (exceptional violation). It is rare that violations are malicious, and people who violate do not normally mean to cause harm.

BOX 15.1
_____

The strength of the human in the system

Humans typically make up for the deficiencies and oversights in the design so that the system actually works in practice. It is human adaptability and flexibility that ensures a functioning and safer system. A classic example of human strength is described in Box 15.1. On January 15, 2009, US airways Flight 1549 took off from La Guardia airport near New York with 155 people on board, destined for Seattle. Within just 3 minutes there was a bird strike from a flock of Canada geese at an altitude of 2818 ft. Both engines failed and soon after forced the aircraft into a gliding descent. Captain Sullenberger and his first officer, Skiles, had precious little time to make critical decisions about the safest way to manage the unfolding catastrophe. They determined that there was insufficient thrust in the engines to reach La Guardia or the alternative Teterboro airport, but that they must avoid crashing in the densely populated city. The decision was taken to land on the Hudson river, and about 90 seconds after the mayday call, the plane made an unpowered ditch, resulting in no fatalities. In this event, the engineered system had failed and there was no safety net to avoid a catastrophe. It was the human element that made up for these deficiencies.

Errors occur due to cognitive malfunction. Someone might make a mistake, that is, do the wrong thing believing that it is right because they have applied the wrong rule to a situation (rule-based mistake) or had a lack of knowledge or understanding of the situation (knowledge-based mistake). Alternatively, it may be a skill-based error, where someone performs a well-practiced task, but not as planned (slip of action) or where they forget a step (lapse of memory). This is illustrated in Fig. 15.1.

Humans fail on a day-to-day basis, but most failures are inconsequential. However, when the human is operating or maintaining a process plant, there is a need to understand how someone could fail. Where the failure could lead to a major accident, the task is referred to as safety critical.

The role of the human factors analyst is not just about identifying how someone could fail, but also what could make the failure more likely.

When analyzing violations, there are typically antecedents (or triggers) for the unsafe behavior, such as inadequate training or peer pressure, but the consequences of the behavior will drive (or motivate) the person to violate. ABC analysis is often used to understand violating

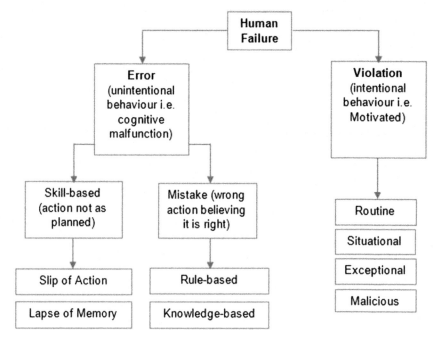

FIGURE 15.1   Understanding human failure.

behaviors (B), taking account of antecedents (A) and consequences (C). A case study is described in Box 15.2.

Errors are more likely if the work conditions are inadequate. These are typically referred to as performance shaping factors (PSFs) or performance influencing factors (PIFs) and include several categories, as described in Table 15.1.

An example of human error during the Texas City accident in 2005 is described in Box 15.3, accepting that there were also procedural violations. In any incident or accident, there will be more than one unsafe behavior, and some of these will be errors and others will be violations.

This explains how humans can fail to achieve the performance expected of them, or indeed the perfect performance that is often assumed. Unless it is clearly understood what impacts human behavior and performance, there is a risk that the engineered system will set people up to fail.

Humans are an integral part of the system and a system is never just "technical." Human failure is systematically linked to the tasks that people are asked to perform; the tools, equipment and workspaces they use; and the environmental conditions and the organizational arrangements in place.

BOX 15.2

ABC analysis being used to tackle an unsafe behavior

Contractors on an oil refinery were frequently being caught (and dismissed) for using their mobile phones while on site. The Keil Center was asked to help understand why this was happening so that measures could be taken to prevent recurrence. A review of the antecedents revealed that there was no safe storage for mobile phones at the gatehouse; there was no signage to remind contractors, they were unaware of previous contractors being dismissed; and the rules were not being reinforced by the security guards. Contractors needed to maintain verbal contact with their companies and site personnel and had no other means of communication other than walking half a mile back to the gatehouse to use the landline phone. The key consequence driving the unsafe behavior was the ability to make calls which were a part of their working requirement. Several actions were taken to tackle each of the antecedents, but contractors were also provided with intrinsically safe phones for use on site. The contractors were then able to achieve what they needed to achieve without breaking the rules, and there were no further repeat incidents.

TABLE 15.1 Examples of performance shaping factors.

PSF categories	Examples of PSFs
Task factors	Task complexity, high workload, task conflict
Procedural factors	Poorly presented instructions, ambiguous wording, an absence of instruction
Communication factors	Lack of communication protocol, high background noise, lack of feedback
Environmental factors	Adverse weather conditions, noise, vibration, poor lighting, thermal discomfort
Training	Poor training and mentoring quality, lack of opportunity for practice, lack of competence assessment
Human interfaces	Inadequate workspace, poor information display through control panels and software graphics, poor tool design
Personal factors	Fatigue, bereavement, lack of confidence, complacency
Social and team factors	Poor supervision, poor shift patterns, lack of inter-team coordination

---

BOX 15.3

Human error during the Texas City accident

Following maintenance activity on the Texas City oil refinery, operators were undertaking a process start up. Part of the process called for filling the isomerization (isom) column to a level of six-and-a-half feet. Operators intentionally filled the column above the six-and-a-half-feet level due to concerns over causing damage to the burners (violation). However, subsequently the column was filled further to a point where it overflowed to the blowdown drum and process sewer and then spilled out of the top of the blowdown drum. The vapor cloud was ignited by a nearby truck backfiring, which caused a massive explosion and multiple fatalities. The error was the further overfilling of the column. The contributing PSFs were the high and complex workload of the board operator; the conflicting instructions on process routing; the lack of shift handovers and log book communications; the lack of knowledge about the dangers of overfilling the column and the lack of simulator training; the flows into and out of the isom being on different control system graphics (and no level indication); and the extreme fatigue and lack of supervisory oversight.

15.3 Gaps within process safety management assurance

The primary weakness associated with PSM assurance, from a human factors perspective, is the failure to adequately understand and assure the human elements within the system. This undermines the ability to adequately assure process safety (because it is a sociotechnical system, not a technical system) and it is part of the reason why accidents still occur.

This does not detract from the enormous improvements that have been made as a result of process safety interventions in developing inherently safer designs.

In the Texas City accident, there were high-level alarms and pressure relief valves, but if the isom column had had a level trip system, operators could not have overfilled the column. If the blowdown drum had had a flare stack, the vapor could have been safely burned off, avoiding the resulting explosion. The engineered protections, had they been present to today's standard, would therefore likely have prevented the overfilling of the isom column and the resulting explosion. They would, however, not have prevented the unsafe behaviors (the errors and the violations). The operators could still have placed the feed valve into manual to raise the level in the isom column, and high-level alarms could still have been ignored.

The technical aspects of the problem appear to have been controlled through engineering, but the behavioral problems associated with process control have not been addressed. The danger is that there is a heavy reliance on those engineered controls. These controls are not fallible, as they too are affected by human involvement. Someone must install a trip system, function test it, calibrate it and otherwise maintain it. Errors can creep into each of these activities, such as failing to calibrate it correctly, function testing the wrong trip system, or failing to reinstate it. Likewise, a flare system must be maintained, and function tested, and errors can creep in to undermine its performance, meaning that it fails to operate on demand. One solution could be to "fully" automate a system, taking the human out of the system. This too provides false comfort if the human element has not been sufficiently assured. There will always be a human, even if the system is "just" maintained by humans.

The focus of attention would be well directed at preventing the human failure in the first place (through task and engineering design). If it cannot be fully prevented, then there is a need to enable detection and recovery from the failure. If it cannot be recovered, controls need to be present to provide the final safety net. Ideally, the engineered control systems, which act as the safety net, will never be called upon, because the human interaction with the system has been well designed and human performance optimized.

The gap that needs to be filled is to fully analyze, assess and audit the human element within the design of the system: What tasks are they being asked to perform? How will they achieve these tasks? How will failures manifest themselves? What will make failures more likely? What are the vulnerabilities if tasks are incorrectly performed? How do we prevent, detect, recover and safeguard against human failures? Are these controls and safeguards operating as they should be?

The time to start asking these questions is during the engineering design stage, but they are equally relevant to be asked during the operational stage of a system life cycle. The main difference is that the opportunity to "design out" human failure is greater the earlier these questions are asked.

15.4 How to assure human performance

Just like PSM, human factors is not just a single assessment or audit. It is a process that starts at the beginning of the engineering life cycle and ends once the system has been decommissioned. Human factors integration (HFI) is a design management activity specifically related to the process of applying human factors within the engineering project life cycle. Further information can be found in the International Oil and Gas Producers guidance [1].

HFI embraces a human-centered design (HCD) approach involving four broad stages (which are iterated throughout the design stages):

- Specify the context of use: What is the human involvement? What form does it take? What are the risks? What are the human-related needs?
- Human factors requirements: Outline what the design needs to meet and how this should be achieved, drawing from the context of use analyses and requirements from legislation, standards, and guidance.
- Human factors in design: Develop the human interfaces to meet the human factors requirements and resolve human factors design trade-offs.
- Human factors evaluation of design: Ensure that the design meets the human factors requirements and end-user needs.

This is shown in Fig. 15.2.

During the detailed design stage and beyond, the human factors involvement continues to cycle through this process. As the design matures, design input and evaluation happen at a greater level of detail until the human factors requirements are satisfied. The broad HFI objectives at each stage of design are shown in Fig. 15.3 and the

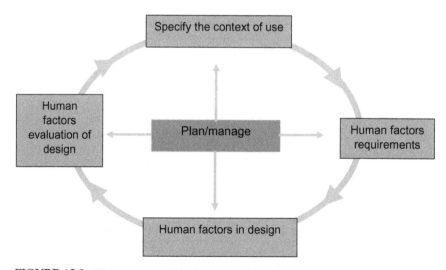

FIGURE 15.2   Human-centered design approach.

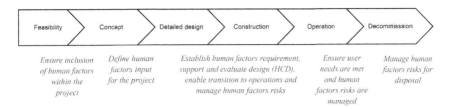

| Feasibility | Concept | Detailed design | Construction | Operation | Decommisssion |

Ensure inclusion of human factors within the project | *Define human factors input for the project* | *Establish human factors requirement, support and evaluate design (HCD), enable transition to operations and manage human factors risks* | | *Ensure user needs are met and human factors risks are managed* | *Manage human factors risks for disposal*

FIGURE 15.3 HFI objectives by life cycle stage.

output from each stage will directly contribute to the assurance of safety.

It is not just about avoiding human failure, but also about optimizing human performance. A relevant example is the control room operator's human machine interface (HMI). The HMI needs to be designed to maintain the operator's situational awareness of the process. This involves effective detection of process deviations, understanding the nature of the deviation and how to handle it, and projecting future states. The operator is thereby set up to predict and manage the process proactively rather than reactively, thus maintaining the plant within acceptable boundaries and avoiding process upsets (and emergency scenarios). This is the philosophy behind the abnormal situation management (ASM) consortium guidelines for effective console operator HMI design, Reference [2]. Again, it comes back to avoiding reliance on alarms and trip systems to counteract suboptimal performance, instead focusing on optimizing performance and avoiding failures and upsets in the first place.

Many process plants around the world have not had this level of comprehensive human factors involvement through the engineering design stages. By the time the system is operational, there is greater reliance on operators and engineering modifications to address any human factors deficiencies. Similar analysis techniques to those used during design can also be used during operations to understand and address human factors risks. A key method is safety critical task analysis (SCTA), which in practice involves three stages:

- safety critical task identification (SCTI);
- task analysis; and
- human reliability assessment (HRA).

SCTI is used to rank order tasks in terms of their safety criticality, that is, their potential to lead to a major accident. There are various methods for performing SCTI and these are described in the Energy Institute's guidance for SCTA, Reference [3]. High priority safety critical tasks (SCTs) are then individually analyzed in detail to determine how the tasks are (and/or should be) undertaken.

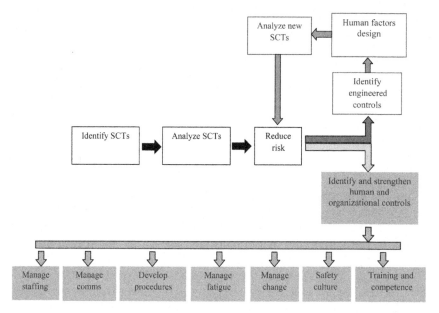

FIGURE 15.4   Assuring human performance.

The tasks within the analysis are subjected to HRA which involves identifying the following for each subtask:
- PSFs that could make error more likely;
- how the error (or violation) could occur;
- the consequences of the failure;
- how the failure could be recovered, if at all;
- the risk controls in place; and
- additional risk controls required.

The additional controls aim to prevent the human failure in the first place (through task and engineering design). If it cannot be fully prevented, then there is a need to enable detection and recovery from the failure. If it cannot be recovered, engineered (or human/operational) controls need to be identified to provide the final safety net.

It is then necessary to ensure that the human and organizational controls are effective, for example, having well-designed procedures, effective training and competence assessment, optimized staffing levels and workload, effective communication protocols, good shift patterns and fatigue management, and a good safety culture.

A model for assuring human performance is presented in Fig. 15.4 which draws together the human factors relevant to reducing the risk of major accidents. Further guidance can be found in the UK Health and Safety Executive (HSE) Human Factors Delivery Guide for Control of Major Accident Hazards (COMAH) [4].

15.5 Concluding remarks

It is misguided to identify the human as the weak part of the system because human performance is only as good as the design of the system. Poor design, which does not take account of human capabilities and limitations, will eventually lead to human failure, whether it is an error or a violation. The weakness in any system is not the human *per se*, rather the lack of suitable attention on assuring human performance during design and operation.

Human factors approaches use structured and comprehensive approaches to understand the human interactions, identify risks where human failure could lead to a process safety incident, and reduce those risks.

Within any auditing process, there needs to be a focus on whether the human elements have been sufficiently covered and whether the process for doing so has been equally systematic as the engineered aspects of the system.

References

[1] International Oil and Gas Producers, Report 454 – Human Factors Engineering in Projects. OGP454, 2011.
[2] ASM Consortium Guidelines, Effective Console Operator HMI Design, 2015.
[3] Energy Institute, Guidance on Human Factors Safety Critical Task Analysis, 2011.
[4] HSE, COMAH Competent Authority - Inspecting Human Factors at COMAH Establishments (Operational Delivery Guide). Version 1.0 April 2016.

Regulating PSM and the impact of effectiveness

Waddah S. Ghanem Al Hashmi[1,2] and Hirak Dutta[3]

[1]Emirates National Oil Company (ENOC), Dubai, United Arab Emirates,
[2]Energy Institute-Middle East (EI-ME), Dubai, United Arab Emirates,
[3]Formerly with Indian Oil Corporation Limited; Oil Industry
SafetyDirectorate. (OSID); Ministry of Petroleum & NG; Nayara Energy
Limited; Indian Society of HSE Porfessionals, Uttar Pradesh, India

16.1 Introduction

Much of the drive toward the implementation of structured management systems was an evolution in the industry and was based on the required compliance of organizations to environment, health and safety (EHS) rules and regulations. There are two types of systems generally that operate to regulate EHS, one which is most prevalent is the prescriptive regulatory systems which operates extensively throughout the world. This is common in many high-risk industries such as construction, maritime and aviation. The process industries also have a great deal of regulation but in certain jurisdictions around the world a more goal-based or performance-based regulatory framework exits, for example, in the United Kingdom and more recently Singapore, etc.

Both systems have their pros and cons, but what is very important to appreciate is that both regulatory type systems require commitment to comply and a management system framework to exist. In the process industries, logically the process safety management system lends itself as an effective framework to help govern compliance.

In this short chapter, we discuss the impact of the compliance verses performance-based approach to process safety management (PSM) and how that may impact on the human element in the implementation of systems. It is not the purpose of this chapter to argue which regulatory framework is better or worse, but just to highlight how regulations can affect performance.

16.2 Purpose of regulations

Regulations provide governments and authorities with an effective reference to ensure compliance. It is assumed that the regulations provide for a minimum safety standard for all industries, thereby, protecting society on untoward harm, protecting workers and those having business with a company which operates, protecting the environment at large, and so on. As such the regulations are purposeful in their existence for all to comply with and as a reference point to ensure that should there be non-compliance for the jurisdiction to deal with fairly.

The development as such of many of the regulations has significant economic impact on operations. Especially starting from the design phase of developing a plant, the regulations generally concern themselves with an expectation of the application of minimum inherently safer design including operating envelops, safety factors, interlocking and so on. Regulations also cover the expectations that all the operations are undertaken within a controlled environment, with competent operators who are trained in both the operations and emergency control procedures. Plants and assets must be kept safe and therefore technical asset integrity management is also an expectation in most regulations.

However, given that most assets and plants are built to specifications, yet have their own specificities and differences based on the front-end engineering design, the regulations are expectation-based and can only be so specific. It is an expectation that in most regulations that the owner and the operator will operate the plant within the operating envelop and must therefore maintain the operating limits.

When considering human factors in the regulations, it is a challenging task, as standards are usually based on design, operating, and technical prescriptive standards. In more recent years to address better the human element in operating and maintaining a plant the regulators have focused more on the competency assurance through reference to the standards that certain independent bodies on standardizing accredited and approved vocational educational programs. In saying this, apprenticeship type training and development has been around for many years.

However, as would be appreciated, especially from the discussions in Chapter 5, **Organisational learning and competency assurance** and Chapter 15, Human factors and PSM compliance assurance activities, competency assurance and human factors is a very challenging and complex subject with different dynamics, which makes it difficult to regulate effectively.

16.3 Prescriptive and performance-based regulations

As discussed, the prescriptive regulatory regiems where developed intially based on establishing the right level of controls in several standardized industries such as the nuclear, power and utilities, aviation and maritime industries. Such industries not only are very standard in their nature, they operate and can have a significant impact on a large number of persons or on large geographical areas. It was therefore necessary that there was to be standardization of regulations. Furthermore, some are required and some even governed by international conventions such as for example the MARPOL or IATA conventions, etc., in the marine and aviation sectors.

The issue in mainly the process industries with some of the prescriptive regulations is that they struggle to manage the dynamic nature of the industry and its risks. Also, the more prescriptive standards may not be able due to the long-cycle for development, review and promulgation to keep up to date with the fast changes in the technology and other developments such as new designs, materials, technology, etc. So, such prescriptive regulations would otherwise stifle growth and development of the industry. Within the aviation industry, the development of the technology is done in a very close working relationship with the regulators and the use of safety cases has been developed over the years. The industry has a much greater collaborative approach and the manufacturers of aircraft and their systems are limited, unlike for example, the oil and gas sector.

But the prescriptive regulations also struggle to deal with the human element aspects, and as explained above they relied more on the regulating of competency development and assurance frameworks based on the assumption of course that human error would occur due to a lacking competency or mistake based on lacking understanding/skill. Of course, as has been discussed in this book, and especially in Chapters 2, 4, and 5, errors occur even with competent operators, and there is a complex set of underlying factors, which regulations would struggle to regulate.

The benefit of performance-based regulations is that it forces the thinking process when developing a process or operation. Right from the early phases with the inherently safe design, the hazard and opera-bility studies and reviews and the dynamic risk assessments creates a better awareness of the potential and credible scenarios in context. This can create a better understanding of the risks and the risk controls and thus when designing the operating procedures, the addressing of the tight coupling between systems can lead to more effective and dynamic risk management and control.

There is another very important aspect as well, it is the accountability and "duty holder" principle, which places greater accountability on leadership to look closely at the operations, the risks and people's beha-viors within the context of the development of a safety culture. This aspect has become better appreciated over the past 1−2 decades in the wake of incidents such as the *Texas BP Refinery Explosion, USA (2005)* [1] and the Baker Panel Report and the *Buncefield Terminal Explosion and Fire, UK (2005)* [2], and the Major Incident Investigation Report (MIIR) that followed. Both the reports placed great emphasis on the role of leadership in driving a safety culture, citing it as being a major aspect that prevents major incidents. In fact, in modern oil and gas related major incidents failing leadership was often cited.

The general trend in modern regulations seems to be moving towards more performance-based regulations globally, especially in the process industries such as oil and gas. Human performance and factors will always be very difficult to regulate. It is the drive toward a better understanding of human factors, addressing human performance issues in the design and risk and operational studies that can shape a better and safer industry in the future.

16.4 Impact of effectiveness of PSM regulations

Regulations generally do not drive performance. They create a stan-dard for industry to comply with in the best interest of the company and all its stakeholders which includes the Government itself and the public. However, failing to comply with regulations can bring about consequences including abatement orders and fines leading to even part or full closure. Therefore, we generally see compliance to the regula-tions as a minimum standard in most industries.

The drive of performance toward more value-adding EHS or PSM comes more from the proactivity and operational excellence drivers. This creates a high reliability culture and organization which leads to a proactive approach toward risk management. High reliability organiza-tions and those who can show safer performance over a longer period create for an organization both economic and other benefits such as higher profitability, lower insurance premiums, less fines, less plant and operational disruptions and can also through creating this better organization, more engaged and loyal employees [3].

Human behavior is significantly affected by the organizational and safety culture within the company. It is therefore recommended to have better systems in place which engage employees, provide the right level of support through training, resources, equipment and empowerment etc. which in turn helps provide a more positive working culture.

Regulations are the minimum requirements that must be complied with—no compromise. Even within a safety case regime, the requirements or the processes and procedures which are created, must be adhered to. Compliance should be a basis, but even with human factors and behaviors, there needs to be an aspiration toward a more proactive and a generative safety culture. This creates a better and higher performance culture which maintains a PSM culture which is more positive and where the employees find that PSM and Safety is a core organizational value rather than being only a legal requirement.

To this end, once more it is leadership that can drive a better safety and PSM culture. When leaders can inspire the employees, an organizational culture toward proactivity rather than compliance can be created in which the employees will be motivated to go beyond compliance and toward a commitment. This sense of personal responsibility and accountability will have a positive impact on their behavior.

16.5 Challenges of having too many regulators

Operators in many developed and developing countries are also faced with the pangs of too many regulators each sermonizing to follow their prescriptions and systems. The concern of too many each vying to grab space and control the industry adversely impacts safety and thereby the business. And the enterprises at times are at sea: whom to follow, what to follow, and what not to follow.

Ideally speaking, there must be one regulator who is equipped with sufficient expertise in preparing safety standards, carrying out audits, identifying gaps in the system and making recommendations for improvements. Such a body must essentially have the requisite knowledge and capability in carrying out Incident Investigations, finding out the root cause of the incident, and recommending measures to avoid recurrence.

It is, therefore, obvious that the regulators must not only have the experience but must have sound domain knowledge too. Industries like that of aviation and nuclear power are globally managed by such regulators who have long experience and expertise backed up by domain knowledge. Similarly, the complex operations in the hydrocarbon industry and its complex processes must also be managed by regulator who fulfill the earlier criteria. It is impossible to do proper justice unless the regulatory body has enough technical expertise and deep understanding of the subject.

In the United States, the Chemical Safety Board (CSB) are called upon by Industry(s) to undertake Accident-Incident Investigation and come out with lessons learned and recommendations to avoid recurrence. The American Petroleum Institute (API) prepares the safety standards which are globally accepted, conducts external audit of the industry to identify the shortcomings and suggesting mitigation measures thereof. Normally, API audits are extensive and rigorous. A team of five to seven experts carries out audit on various process safety parameters over a period of 5 days. These organizations, although not necessarily regulators, have the requisite knowledge, expertise backed up by several years of experience and can deal with the complex issues in a professional manner.

Organizations like API and CSB are effective because they not only depend solely on their own internal resources but regularly draw experts from relevant industry. The confluence of industry professionals and the expertise of Institutes confirms complete focus on arriving at solution rather than their propagating their own self-identity. It is a win-win proposition.

However, such narratives are not profound around the world. In India for example, and in safety in oil and gas industries, in a third-largest consumer of energy and the fourth largest refiner in the world, is managed by multiple regulators. There are four major regulators overseeing safety in the hydrocarbon sector. This leads to unnecessary intrusions in the industry like multiple number of audits, complying with multiple prescriptions and suggestions, investigation by too many agencies, and so forth.

This happens because there are many areas of overlap since the role of each of the Regulators are not clearly delineated. Thus, industry must face the wrath of multiple regulators which is neither desirable nor healthy. Too many audits by multiple bodies and interferences thereof hinders the productivity of organization and some recommendations at times can contradict one another. There are such similar examples in many other countries and jurisdictions too.

16.6 Chapter concluding remarks

In this short chapter, we have discussed the role of regulations in driving PSM compliance and performance. The chapter briefly discussed the difference between prescriptive and performance-based regulatory frameworks and spoke of some of the pros and cons especially in the context of human factors. It is to be appreciated that in the process industries sector, regulations have moved more toward performance-based standards and regulations due to the dynamic and tightly coupled systems and associated risks.

Regulations provide a minimum set of standards that must be complied with, whereas more performance-based safety case led regimes look at the risks and try to address risks in a pragmatic approach. PSM systems within high reliability organizational cultures rely on leaders to drive that sense of personal responsibility and accountability within the workforce which has very positive impacts on the human element in the sense of performance. In closing, this chapter also spoke about the challenges of having too many regulators, the advantages of having semiregulatory and industry bodies helping in setting the right standards of practice for the processing industries as well.

References

[1] J. Baker (III), F.L. Bowman, E. Glen, S. Gorton, D. Hundershot, N. Levison, S. Priest, T. P. Rosentel, D. Wiegmann, D. Wilson, The Report of the B.P. U.S. Refineries Independent Safety Review Panel, January 2007. Available on-line from http://www.csb.gov; 2007.

[2] Major Incident Investigation Board (MIIB), The Buncefield Incident 11 December 2005: The final report of the Major Incident Investigation Board. Published by the Control of Major Accident Hazards (COMAH) and see http://buncefieldinvestigation.gov.uk, 2011.

[3] W. Al Hashmi, Environment, Health and Safety (EHS) Governance and Leadership – and the making of High Reliability Organizations, Routledge, United Kingdom, 2018.

17

Readying the organization for change: communication and alignment

Chitram Lutchman and Ramakrishna Akula

Safety Erudite Inc., Canada

17.1 Introduction

In this chapter, the authors discuss how to prepare an organization for the implementation and alignment of the process safety management (PSM) across the organization. The authors address the importance of understanding how to transition PSM from immature to sustainable states within the company's operating systems and processes. The authors also address the organizational change management requirements necessary for stakeholder alignment, buy-in, and commitment necessary for implementing PSM across the organization. The authors highlight the importance of a stakeholder impact assessment and the need for adoption of appropriate strategies necessary for effective communication and adoption of PSM across the business. The goal of the authors is to ensure that leaders are adequately prepared for the hard work, which is referred to as the "heavy lift" associated with effective alignment and communication required for the successful implementation of PSM across the organization.

17.2 Key elements of organizational readiness and alignment

Organizational survivability today is dependent on organizations doing more to prevent unplanned events that have the potential to

Process Safety Management and Human Factors.
DOI: https://doi.org/10.1016/B978-0-12-818109-6.00017-4
221

cause injuries to its workforce, protect the environment and eliminate damage to its assets. More importantly, organizations are required to demonstrate genuine care for the environment in their social responsibility programs and solid business performance to attract and retain the best and the brightest [1,2]. Leaders should, therefore, recognize and leverage the values of PSM in its social responsibility programs so that it may continue to attract and retain high quality committed workers. Readying the organization for PSM must be built upon a solid platform that is designed to clearly articulate the following:

- What is the organizational vision for PSM?
- Why is PSM required for the organization, and is the organization committed to full implementation and sustainment?
- What additional benefits does bring to the organization?
- How is PSM different from what we are doing now?
- What is the work required/effort involved in implementing PSM?
- Who is impacted, and how impacted are they?

17.3 Creating a shared PSM vision

Successful launch and implementation of PSM require a shared vision for PSM across the organization. Leaders must articulate clearly and concisely what PSM brings to the organization as an ultimate outcome for the organization. The vision must be communicated with a cadence that drives all members of the organization to want to become a part of the movement toward this vision. What this means is the PSM vision must meet the criteria defined below. It must:

- Be
  - Believable
  - Achievable
  - Realistic
  - Easy to share
  - Simple and easy to recall
- Resonate with all levels of the organization
- Become like a jingle in people's minds: creates a cadence

17.4 Sharing the vision

Sharing the vision is the outcome of genuine belief in and communication of the vision to all levels of the organization. Senior leaders of the organization must get to all levels of the organization and through to the frontline, and communicate the vision to workers. Sharing the vision requires clear identification and communication of what's in it for me (WIFM) for the appropriate stakeholder group. Sharing the vision may require the following:

- Use of multiple communication methods and tools
- Multiple field visits and communication sessions
- Communication at the right frequency levels

The effectiveness of each communication method for sharing the vision varies. The most effective method for sharing, however, involves face-to-face interactions between leaders, workers and stakeholder groups. Communication (reflected in words and body language) must demonstrate genuine belief and commitment to PSM and should be able to create trust in the workforce that we are implementing PSM for the right reasons: preventing loss of primary containment (LOPC) and protection of the workforce, the environment, the assets and the image and goodwill of the organization Kennett-Hensel and Payne [3] explained that sharing of the PSM vision is a function of leadership. Tucker and Russell [4] describe the transformational leader as one who can create and share an inspiring vision, communicate this vision, encourages people to give more than expected and makes pursuing the vision worthwhile for all stakeholders.

17.5 Aligning the organization—organizational change management

What do we mean by organizational change management? Change management refers to the ways people change is managed in the organization. Change management focuses on:

- Identifying the impact of a proposed change on each stakeholder group within the organization
- Developing the mitigation actions required to minimize the impact on each stakeholder group
- Defining the following:
  - Communication messages and communication methods for each stakeholder groups
  - Training needs for each stakeholder group

Haudan [5] advised of eight behavioral ground rules for successful change management. When applied to PSM, they include the following:

- Assume positive intent from the implementation of PSM and trust the experts who are knowledgeable about assigned disciplines

- Communicate with clarity. Clearly communicate the requirements and expectations of PSM for ourselves and our people, and not to stop until everyone is clear
- Make sound decisions regarding the right balance for PSM applicable fact-based decisions with getting it done in the absence of full information: begin implementation where known gaps may exist. Strive for the 80/20 rule in having the right information before acting
- Encourage everyone to think about the organizational benefit vs. my business unit or my department
- Be candid, honest and open in field communication and discussions. Avoid telling one group one thing and another group something different
- Demonstrate commitment to actions, support decisions. It is ok to revisit, but do so with the entire stakeholder group or team
- Demonstrate learning and practice sharing of learning. Avoid concerns of sharing dirty laundry in public. Learn to prevent a repeat of the same or similar incidents
- Communicate frequently on PSM implementation Progress - tell it straight, never lie or seek to mislead the stakeholder group or team - avoid sugar coating

17.6 Conducting the stakeholder impact assessment

In its simplest form, the stakeholder impact assessment is intended to understand the change impact upon various stakeholder groups and levels of the organization. Where PSM implementation is concerned, change impact assessment should be performed at the following organizational levels:

- Senior leaders
- Middle managers
- Frontline Supervisors
- Frontline workers

The impact of PSM is also more closely felt across operating assets than across support and functions. Consequently, therefore, change management associated with PSM implementation should ensure focused attention to various stakeholder groups and support functions across the operating sites. With centralized controls and support of operating assets, change management strategies should be strategic (across multiple sites) where it needs to be and specific (site-specific or local) as applicable in addressing change management requirements and needs.

When performing the stakeholder impact assessment, Table 17.1: Change Impact Focus: Definitions and Examples, provides the key focus

TABLE 17.1 Change impact focus: definitions and examples.

Impact focus	Definition	Example
Processes	Refers to defined approaches to doing some type of work.	Steps applied in managing change: MOC
Systems	The collective integration of people, processes and assets to produce defined organizational outcomes.	The introduction of an Operational Integrity Manual to improve reliability, HSE and business performance
Tool	Tools refer to items or processes used to meet the requirements defined by PSM in any of its respective Element.	A Computerized Incident Management Database (CIMD) A Computerized Maintenance Management System (CMMS) A risk matrix for assessing risk levels
Job Roles	A description of what a worker is required to do in an assigned organization chart box and the competencies required to perform the duties safely, effectively and efficiently.	Engineering Manager / Network Members / Training Specialist / Operations Supervisor etc.
Critical Behaviors	Refers to required behavioral responses by a worker or team to PSM related stimuli	Individual worker and or emergency response team behaviors in response to a site emergency or loss of primary containment during operations of an asset
Mindset /Attitudes /Beliefs	Refers to the worker or stakeholder group general mindset toward the implementation of PSM	Apathy, been there done that when versions of PSM gas been implemented before; Motivated - get it done attitude; Excellence - doing the right thing the right way every time
Reporting Structure	The positional relationships in the organization that defines reporting and hierarchical structure.	Organization structure that defines the reporting structure of the organization - who reports to whom
Performance Reviews	Performance reviews refer to the processes and indicators used to measure progress toward defined PSM goals and objectives	PSM performance indicators such as reliability and HSE performance

(Continued)

TABLE 17.1 (Continued)

Impact focus	Definition	Example
Compensation	Refers to monetary and nonmonetary rewards offered to implementation of PSM.	Refers to pay scale changes for additional work; a safer work environment and improved work-life balance from a PSM driven safer and more reliable operating asset
Location/ Operating Site	Refers to a physical operating site/asset with operating processes to deliver the business products.	A site/department associated with the organization's business processes and production outputs

areas to determine the impact on stakeholder groups. For PSM, therefore, the impact of each PSM Element is assessed through on each stakeholder group to understand and mitigate any adverse impacts of the proposed change. Prosci [6] proposed the various change focus areas defined in the table below that should be considered when performing the stakeholder impact assessment.

Table 17.2: Stakeholder Impact Assessment Template for Each Stakeholder Group, is designed to assist those involved in the implementation of PSM In understanding the magnitude of the changes across various stakeholder groups of the organization. This template helps the organization better understand where the change impact on stakeholder groups is the same or similar and can be addressed using the same methodology or perhaps at the same time.

Where the change impact is moderate and higher, the organization should seek to ensure frequent and effective communication to ensure all stakeholder mitigation actions (risk management, communication and training) are met to improve the likelihood of successful PSM change.

When properly completed, the stakeholder impact assessment provides a platform for aligning the organization for successful PSM implementation. The stakeholder impact assessment should identify for the corporation the following for the various stakeholder groups:

- What is changing
- The impact of the change
- Mitigation requirements and need for:
  - Communication
  - Training
  - Risk management
  - Resourcing
  - Organizational capacity for change
  - Prioritization of work

TABLE 17.2   Stakeholder impact assessment template for each stakeholder group.

Impact focus	What is changing	Change impact			Mitigation actions
		High	Moderate	Low	
1. Processes		☐	☐	☐	
2. Systems		☐	☐	☐	
3. Tools		☐	☐	☐	
4. Job Roles		☐	☐	☐	
5. Critical Behaviors		☐	☐	☐	
6. Mindset/Attitudes/ Beliefs		☐	☐	☐	
7. Reporting Structure		☐	☐	☐	
8. Performance Reviews		☐	☐	☐	
9. Compensation		☐	☐	☐	
10. Location		☐	☐	☐	

Successful change is derived when the mitigation requirements are addressed for each stakeholder group.

17.7 Managing organizational change

Safety Erudite Inc. [7] advised that people and organizational change related to PSM are best achieved when all essential components defined in Fig. 17.1 is adequately addressed for each stakeholder group.

Effective engagement and communication are required to ensure PSM requirements and expectations are adequately communicated and stakeholder risk mitigation actions. It is important, however, to ensure the most effective communication methodologies are adopted and applied to maximize adoption during the implementation and sustainment of PSM.

Factors that may adversely impact PSM Change are presented in Table 17.3. Care and attention are required to ensure that these conditions are properly addressed.

17.8 Communication methodologies

Table 17.4 highlights the pros and cons of different communication methodologies that may be applied by leaders of the organization in communicating PSM. Among the pros and cons are the following:

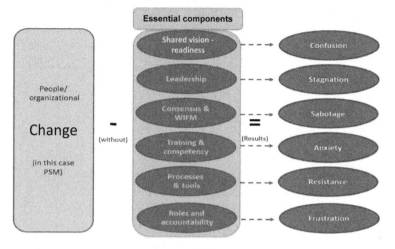

FIGURE 17.1 Components of people/organizational change management. Source: *Safety Erudite Inc, OEMS – Getting It Right the First Time. Unpublished presentation, 2019. Retrieved May 10, 2019 from www.safetyerudite.com.*

17.9 Seven best practices in organizational change management

Prosci [6] pointed to seven best practices for organizational change management. Table 17.5 provides a summary of the implications of each of these best practices.

17.10 Summary

In PSM, like in other management system changes, should follow a structured change management process that integrates change management personnel with the PSM project management team to enhance success and sustainment. Readying and aligning the organization for PSM implementation is a complex process that requires the following:

- Committed leadership
- A completed stakeholder impact assessment that identifies the most impacted groups and risk mitigation plans
- Leveraging the seven best practices for successful change management
- Avoiding the pitfalls that may adversely impact PSM change

TABLE 17.3   Factors impact PSM change.

Factor	Details
Lack of or No Vision	• A poorly communicated vision that is not shared by stakeholders leads to confusion among stakeholders.
Inadequate or no Communication	• Weak or poor communication can lead to key stakeholder groups like the middle management team and frontline supervisors.
Inadequate or no trust	• When the ABCDs of trust is absent in the communication and behaviors of leaders is absent, change generally fails. With PSM, when leaders are not Able and competent, Believable in actions, Connected to the workforce and fail to Demonstrate commitment in behaviors, PSM implementation generally fails.
Ego	• When leadership ego gets in the way, PSM implementation generally fails or is suboptimal.
Fear of the Unknown	• When workers are unaware of the value of PSM, the WIFM and personnel safety and asset integrity are poorly communicated, PSM implementation may fail.
Weak or Reactionary Planning	• Poor or weak planning for the implementation and sustainment will result in failure. Lutchman *et al*, (2019) pointed to the need for Networks for successful implementation of OEMS across multiple sites.
Accountability Failures	• When everyone is accountable, and no one is accountable, PSM implementation will likely fail. Everyone thinks the other is accountable.
Rapidity of Change	• Organizational capacity (too many ongoing activities and projects) and the heavy lift associated with PSM can lead to worker and system overloads. When this occurs, the pace of implementation must be adjusted to prioritized PSM elements and priorities.
Silos	• When organizational silos are developed, poor integration between sites and departments may lead to weak or poor PSM implementation. Lutchman [8] pointed out that Networks can eliminate organizational silos that may impede PSM.
Stakeholder Disregard or Disengagement	• When poor stakeholder impact assessment is performed, or stakeholder impact mitigation is not addressed, the people side of change associated with PSM is adversely impacted.
Corporate/Departmental Politics	• Budgetary constraints, allocation based on size and bargaining powers of leaders introduce politics in the decision-making process. PSM implementation and resource allocation should be risk-based for effective implementation and protection of the organization.

TABLE 17.4 Pros and cons of PSM communication methods.

Communication method	Pros	Cons
Face-to-face	• Very effective and personal: small groups generally communicated with • Easily communicates the what's in it for me (WIFM) • Can easily answer the questions of stakeholder groups • Can recognize and address body language	• Costly and time-consuming • Logistically difficult to plan when multiple plants and sites are involved • Leaders can experience burn out travelling from site to site • Not all leaders can communicate with conviction and commitment • Difficult to communicate with large audiences
Print and Graphic Media	• Easy to get to large audiences and stakeholder groups at the same time • Easy to produce • Very cost effective	• Impersonal and without emotion • Body language impact is not visible • Readers interpret information differently so key messages may be lost in transit
Electronic	• Easy to get to large audiences and stakeholder groups at the same time • Very cost effective • Can easily receive feedback from stakeholder groups • Questions and concerns from stakeholder groups can be easily addressed • Can be personalized by leaders	• Often impersonal and emotion may not be clearly communicated • Body language impact is not visible • Readers interpret information differently so key messages may be lost in transit • Poor messaging and errors can be easily forwarded and shared to others
Website Information	• Useful in communicating updates on PSM implementation progress • Can be used to share success stories • Can provide continuous access to corporate and site PSM information	• Depends on a pull process to get stakeholders to access the site • Impersonal and without emotion • Body language impact is not visible • Readers interpret information differently so key messages may be lost in transit

TABLE 17.5 Seven best practices for organizational change management.

Best practices	Details
Assign an active and highly visible primary PSM sponsor	PSM change management is most affected by an active and highly visible primary sponsor who: • Provides consistent attention to the change and the need for change management in PSM implementation • Leads and motivates others in the organization • Makes effective and influential decisions including aligning priorities across the organization • Maintaining direct involvement with both the PSM implementation team and the change management resource throughout the implementation process
Include a dedicated change management resources	A dedicated change management resource is essential to support PSM implementation as follows: • Funded and adequately resourced • Experienced and competent in change management
Follow a structured approach to change management	PSM implementation shall follow a change management process that is: • Defined and communicated • Customizable and scalable to each site • Easily implemented across multiple changes and all stages of the project
Engage and involve employees	Engage and involve employees as follows: • Highlight the WIFM for each stakeholder group: make it personal ... safety, reliability and business performance • Mitigate risks associated with impacted groups and stakeholders
Communicate frequently, frankly and openly	PSM implementation team should communicate as follows: • With cadence: strong adherence to the key messages • Consistency: strong adherence to the key messages • Transparency: Honesty - telling it straight and never lying • Leverage most appropriate communication channels as defined earlier

(Continued)

TABLE 17.5   (Continued)

Best practices	Details
Integrate change management with PSM project management	Integrating change management with PSM project implementation is achieved by: • Including change management deliverables in the project plan: for example, the stakeholder impact assessment • Participate as a member of the project team • Delivering change management training to the PSM implementation project team

Engage and involve middle managers and frontline supervisors	Engaging and involving middle managers and frontline supervisors is achieved as follows:
	
	• Ensuring managers deliver PSM communication to department and followers
	• Ensuring managers and frontline supervisors:
	  ○ Communicate with teams
	  ○ Build PSM awareness that includes how PSM will affect them, the business reasons for the change, and the need for change management
	  ○ Are provided knowledge and expertise to help them meet PSM expectations
	  ○ Are engaged throughout the PSM implementation and sustainment lifecycle

In conclusion, change management is fundamental to the success of all projects. When applied properly in addressing the people side of change, the likelihood of success is improved considerably.

References

[1] L. Leemen, C. Li-Fei, Corp. Soc. Responsib. Environ. Manage. 25 (5) (2018). Retrieved May 10, 2019 from GreenFILE Database.
[2] R.E. Slack, S. Corlett, R. Morris, Exploring Employee Engagement with (Corporate) Social Responsibility: A Social Exchange Perspective on Organisational Participation. Retrieved May 10, 2019 from Education Source database, 2014.
[3] P.A. Kennett-Hensel, D.M. Payne, Guiding principles for ethical change management, J. Bus. Manag. 24 (2) (2018) 19—45. Retrieved May 13, 2019 from EBSCOHost Database.
[4] B.A. Tucker, R.F. Russell, The influence of the transformational leader, J. Leadersh. Organ. Stud. 10 (4) (2004) 103—111. Retrieved May 03, 2019 from EBSCOHost Database.
[5] J. Haudan, Plan for successful change management: keep in mind these 8 core behaviors, Leadersh. Excell. 36 (2) (2019) 37—39. Retrieved May 13, 2019 from EBSCOHost Database.

[6] Prosci, 2019. The Best Practices in Change Management. Retrieved May 10, 2019 from https://www.prosci.com/resources/articles/importance-and-role-of-executive-sponsor.

[7] Safety Erudite Inc, OEMS – Getting It Right the First Time. Unpublished presentation, 2019. Retrieved May 10, 2019 from www.safetyerudite.com.

[8] C. Lutchman, C. Lyons, K. Lutchman, R. Akula, W. Ghanem, OEMS- Getting It Right the First Time, Taylor and Francis Publication, 2019.

Do we really learn from loss incidents?

Ram K. Goyal

Bahrain Petroleum Company (BAPCO), Bahrain Refinery, Southern
Governorate, Bahrain

18.1 Introduction

The general public's perception is that those in the chemical, petrochemical, oil, and gas industries mostly do not learn well from loss incidents of the past and that is why we keep on having repeat incidents. In the past decade, we have had several major incidents that resulted in massive financial loss and property damage and/or multiple fatalities in refineries and gas handling facilities in Europe, the United States, and the Middle East. Some of the "business interruption" insurance claims have been so large that some insurers are seriously considering withdrawing from the refinery and construction classes of business.

Nonetheless, there are many organizations and individuals who take this learning from incidents very seriously. We can draw very useful lessons not only from major incidents that have caught the public and media eye, but also from smaller incidents that occur in one's industry. When an incident is reported in the media everyone is naturally curious as to what happened. But if we want to learn from the incident we must go beyond simple curiosity. There are some known hurdles in our ability to learn from other people's incidents. In order to learn lessons, we need to review the incident in depth. In analyzing these incidents for the purpose of learning lessons for the future, there is no point in trying to blame specific person or persons. "Culpability" is applicable in cases of sabotage or other deliberate actions with malice aforethought. In accidents and incidents, it is better to concentrate on systemic or

Process Safety Management and Human Factors.
DOI: https://doi.org/10.1016/B978-0-12-818109-6.00018-6

procedural failures or design deficiencies so that viable, effective and long-term remedial measures can be proposed and implemented.

The learning process is not easy. You need to have the will and the ability to pay attention to detail.

> Safety management is not rocket science. Rocket science is trivial pursuit compared to the management of safety. There are only a limited number of fuel types capable of lifting a payload into space: but the variety of ways in which harm can come to people is legion. —*Trevor Kletz (from his book,* **Still Going Wrong!***) [1]*

The US government has established a Chemical Safety Board (CSB) with a view to conduct detailed investigations of major incidents that occur in the US and which become focus of media and public attention. To date the CSB has conducted several such investigations and have made numerous recommendations aimed at preventing recurrence. The CSB investigation reports are particularly useful to those who are seeking to discover root causes of the incident and not merely the newsworthy aspects of it. In addition to written detailed reports, the CSB sometimes commissions production of accident sequence animations that are a great help in understanding all pertinent details. In a recently issued report, the CSB reported their findings of a major incident that occurred in an oil refinery belonging to a US major oil company.

Case Study 1

The propane release was likely caused by the freeze-related failure of high-pressure piping at a control station that had not been in service for approximately 15 years. The control station was not isolated or freeze-protected but left connected to the process, forming a dead leg. Water in the propane accumulated in the low point formed by the control station and froze during cold weather prior to the incident, cracking an inlet pipe elbow. Ice sealing the failed pipe from the process melted as the temperature rose on the day of the incident, releasing 4,500 pounds per minute of liquid propane, which ignited. [2]

For details of this incident, please refer to the CSB website where an animation video of the incident can also be seen. The above is a clear case where a "dead leg" in the process piping failed. It can be argued that the hazards associated with dead legs in process plants are quite well-known since many other incidents have occurred in the past in our industry. How come then a safety-conscious US major oil company could still suffer a terrible loss in 2007 (multiple serious injuries) due to a well-known and well-understood cause? Are we not learning from our past mistakes? We may be good at learning from our own incidents but perhaps we are not so good at learning from other companies' incidents.

18.2 Barriers to learning

There are some recognizable barriers that we tend to raise. This happens when there is a lack of true learning culture in an organization. Some of these are discussed subsequently.

18.2.1 We already know about it

Sometimes we might fool ourselves into believing that the incident in question is quite familiar and has happened many times and thus does not deserve further attention. Many motor vehicle—related incidents that we read about in the daily newspapers fall into this category. For a few moments, we become concerned about our own driving habits but soon thereafter we go back to our set routines. A case in question is the Buncefield incident. A storage tank at the Buncefield oil depot North of London was overflowing for more than 40 minutes before an explosion occurred in which 43 people were hurt, and nearby homes and businesses destroyed by the fire that followed. It should be noted that overfilling of a tank or a process vessel or a distillation column is not an entirely unknown phenomenon in the oil industry. This has happened many times in many companies and we all know about the possibility. However, if we just ignore Buncefield simply as another incident of tank overfilling, then we are letting a good learning opportunity slip by us because the incident investigation reports contain a host of other valuable lessons that deserve our attention.

18.2.2 The people who had the incident were stupid

One of the contributing factors to the Bhopal tragedy was that one of the Union Carbide managers had decided to switch off the air-conditioning unit that was meant to keep the contents of that tank cool in order to prevent a run-away heating up. He had thought that since it was the month of December, the outside ambient temperature was low enough, so why waste energy in keeping the A/C on? In hindsight, this might seem quite stupid; but Union Carbide did not employ stupid people. In fact, prior to the tragedy, Union Carbide was a highly respected company in India and graduates from well-known technology institutions (such as the Indian Institutes of Technology) used to line-up their corridors seeking employment! No one in his right mind goes to work in the

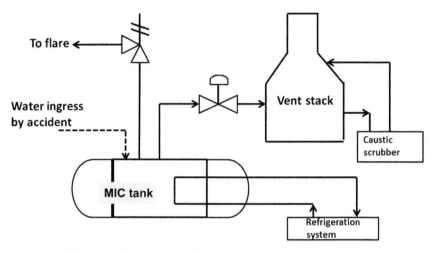

FIGURE 18.1   Bhopal before the incident.

morning thinking that today he would blow-up the plant. We must never ignore the lessons from an incident under this pretext. (Fig. 18.1)

18.2.3 This is not applicable to our situation

This is quite a tough barrier to demolish from our mindset. Since it is likely that no two incident situations will be identical, we can always argue with ourselves that our situation is different; our process parameters are different; we do not have a vessel of this size; our metallurgy is different and so on. If we have a truly learning culture in an organization, then we must not get tangled into the differences between the plant that had the incident and our own situation; we should carefully scrutinize the potential similarities or even somewhat relevant   scenarios that could land us into trouble if not addressed satisfactorily.

For example, consider this incident where a crane toppled over from the transporter that was carrying it. It is certainly a traffic risk; however, it also presents a risk in terms of likelihood of the crane falling over some oil or gas lines, releasing flammable hydrocarbons into the atmosphere which could get ignited and thereafter lead to a fire or explosion incident. The resulting consequences in that case could have been far more severe. Some refinery personnel reviewing this incident might conclude that there are no lessons to be learned from this because their situation is different. However, one could gain some useful pointers from this that would relate to transportation of heavy, awkward objects, an activity that surely takes place in refineries during turnarounds and even sometimes while the units are operational.

18.2.4 Our standards and procedures are better

Sometimes we raise this barrier in the belief that our standards, practices and procedures are better than those adopted in the country or the organization in which the loss incident under consideration has taken place. While it could be true that some of our standards are more stringent or more robust when it comes to process safety management (PSM) principles, but in terms of their application in practice, it is possible that we could also have chinks in our armor. A defective boiler explosion led to further loss of containment of hydrocarbon materials in the Skikda refinery [3].

Flames lighted the sky after an explosion at Algeria's largest refinery and principal oil exporter in the port city of Skikda, 500 km (310 miles) east of the capital Algiers, early January 20, 2004. Rescue workers searched through rubble for missing worker on Tuesday after a huge explosion at its key gas installations killed 23 people on Monday evening. We could argue that our standards ensure that boilers are inspected on a regular basis by an external agency under the laws of our government; and therefore, our standards and inspection processes are better. But this does not mean that we should brush aside the Skikda incident as one which does not have any lessons for us. Details available on this incident provide quite useful and pertinent lessons in how to implement recommendations made by insurance surveyors, how to respond to emergencies, and so on.

18.2.5 We have never had such an incident

"Every serious accident is unique, since each one includes various elements which only occur once"—Gérard Mendel [4]. Some incidents are so bizarre that initially it is difficult to comprehend how the incident could be of value from a lesson-learning standpoint. One case in point was an almost new offshore platform belonging to Petrobras that tilted and capsized within a very short space of time (less than 2 hours) [5].

For organizations that do not have offshore facilities a natural inclination would be to assume that this incident is of little interest to them. Nevertheless, given below are two of the recommendations that were made by the enquiry commission that investigated this incident:

- Improve the definition of responsibilities as they relate to the operation, maintenance and supervision of areas of production, platform infrastructure and control of stability. Examples: Supervision/operation of equipment facilities linked to the processing activities (Waste Oil Tank, EDT); operation of ballast and infrastructural equipment.

UK offshore - over 3-day injury rate (per 100,000 workers)

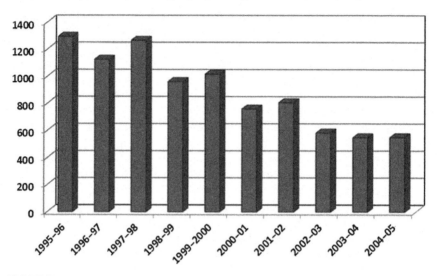

FIGURE 18.2 UK offshore injury. Source: *Data from Health and Safety Executive (HSE) Reports.*

- Review supervisors' functions to reduce their bureaucratic activities and to concentrate their focus on operating activities, such as: operations in progress; non-routine adjustments; restrictions caused by warning or safety features; operational procedures, and the complexity or risk of which requires their personal intervention.

From the above recommendations, even though the incident occurred offshore, it could still have many action items that will be valid for organizations that operate only onshore (Fig. 18.2).

In the United Kingdom, the offshore industry implemented many risk reduction measures following the well-known Piper Alpha disaster with the result that their safety performance steadily improved during the period that followed Piper Alpha (see bar diagram [6]). Nonetheless, there are many other parts of the world where lessons from Piper Alpha and the Petrobras incidents still need to be implemented.

18.2.6 We have already had this type of an incident

This might seem rather incredible but there are those who believe that if you have already had an incident then it is statistically less likely that you would have the same type of incident happen to you shortly thereafter! They draw this conclusion by a misapplication of statistical theory that states that the likelihood of two loss events occurring per unit of time is less than the likelihood of only one event occurring. They fail to recognize that once an incident has occurred, statistical theory is not what will prevent recurrence; implementation of effective remedial measures will.

18.2.7 It just can't happen to us

Consider incidents such as Chernobyl, Challenger space shuttle, Columbia space shuttle, lost Mariner Mars probe, sinking of Harold the Enterprise, fire at Kings Cross tube station in London, the Russian train disaster, and aviation incidents. It could be argued that these types of incidents could not occur in the process chemical or oil and gas industry because the nature of our business is quite different from those organizations that were involved in the incidents exemplified above. However, one must just look at these incidents beyond the newspaper headlines into some of the contributing factors and root causes, it would become quite obvious that there are many general principles and lessons (especially in the area of minimizing human error and making training more effective and relevant) that can be learned from incidents that have occurred in other industries, organizations or countries.

Let us illustrate this point by an example. Many a times we have heard of incidents in Nigeria that have resulted in hundreds of fatalities of people who were trying to fill their cans with diesel or gasoline by puncturing holes in cross country product transfer pipelines belonging to oil companies. The escaping/spilled fluid got ignited in these cases resulting in massive fires that engulfed the people gathered around the leak site. On the face of it one can argue that since this is caused by vandalism, it cannot perhaps happen in our case or in   other economically advanced countries. However, what needs to be highlighted is the fact that another causal factor in the massive number of casualties in these cases was due to a total lack of awareness of the hazard involved on the part of the people participating in this activity. A similar hazard exists when busy road traffic or housing developments are permitted to come too close to our hydrocarbon pipelines or other oils and gas facilities. Likewise, a more worrisome situation exists when people erect their vacation tents close to our lines and oil field facilities.

18.2.8 Knee-jerk reaction, then all quiet

Sometimes when a large loss event occurs, thereby developing a significant public concern fueled by media hype, managers in the related industries might be forced to display what can only be described as a

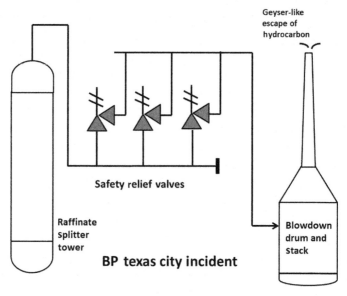

FIGURE 18.3   Schematic diagram of relief system.

knee-jerk reaction. They come up with hastily developed mitigation measures which might not prove to be effective in practice. And gradually when the memory of the incident fades in the public mind and the media pressure eases, all such mitigation plans get set aside because they never thought through properly and were never developed as viable and effective measures. The knee-jerk reaction is only meant to assuage potential public angst, keep the media at bay, and keep the Boardroom quiet.

A case that illustrates this is the now well-known BP Texas City refinery explosion and fire of March 23, 2005 which resulted in 15 fatalities, 180 injuries, and massive damage to the plant as well as to properties outside the refinery fence (windows of some houses located as far as 8 km away reported to be broken by the blast) (refer to Fig. 18.3).

When news of this incident and some rumors about the way it happened were aired by the media, several companies displayed their knee-jerk reactions to it. The hazards associated with on-plot relieving of hydrocarbons (instead of flaring at a safe distance) had been well-known in the industry even prior to the BP Texas City incident. Yet many companies started conducting in-house reviews to determine if they had any facilities that relieved hydrocarbons through vents instead of burning those in flares. Surely, such reviews should have been conducted by these companies without waiting for a major incident to happen somewhere.

Knee-jerk reaction by itself is not a bad thing. The problem is that reacting in such a hurry we might not make the right choices for our company; and secondly, our efforts get deflated with the passage of time. The on-plot relieving of hydrocarbons was just one factor in the incident causation. The CSB investigation report demonstrated that there were many lessons that the industry needed to learn from this incident. One notable one was the development of a whole new approach to siting of temporary trailers and portacabins to accommodate the large workforce (contractors as well as company employees) that is deployed to carry out turnarounds and inspections activities during periodic shutdowns and start-ups of process units.

18.2.9 New incident grabs attention

When a major incident occurs following another one that is still fresh in the public mind and is still being covered in the media; the new incident can grab the limelight and people and media forget the previous one. In this way, any lessons from the previous one can get diluted because of the diverted attention.

A case in point is the PEMEX Mexico City LPG major incident. On November 19, 1984, there was a massive fire and explosion incident at this LPG storage terminal of the national oil company of Mexico (PEMEX). The incident resulted in boiling liquid expanding vapor explosion (BLEVE) of several LPG vessels and bullets which destroyed the townships surrounding the facility. More than 500 people were fatally burned, and some additional 6000 people had burn injuries of various severities. The incident caused somewhat of an uproar in the industry because hazards related to LPG had been earlier highlighted in an incident in France in the 1960s (Feyzin). However, a couple of weeks later, on the night of December 2, 1984 (and early morning of December 3) the Bhopal tragedy occurred in India and during the months that followed PEMEX Mexico City was virtually forgotten and remained off the radar of media and public attention.

18.2.10 People get moved/promoted

Any major incident is likely to have a profound and long-lasting impact on the safety attitude and behavior of the people who are involved in it or who have witnessed it from close quarters. However, as time goes by, the people who had learned the most from the incident move on or leave the company or get promoted to higher ranking jobs. Therefore, it is extremely important that lessons learned from an incident result in sustainable improvements in safety and loss prevention systems. For example, if the lesson from a fire incident was that we need to provide better training to our fire fighters then it is not sufficient to do so for say a year or two; we must make sure that we devise a system that delivers enhanced (and improved) training of fire fighters on a continual basis.

18.3 How to learn more effectively from external incidents

In general, most companies have adopted formal procedures and systems to deal with incidents that occur within their facilities or within their jurisdictions. These might include a notification procedure, an incident investigation system, and an action items implementation and follow-up process. What most companies do not have is a formal system of learning from other people's incidents; incidents that are reported in the press and investigated by independent third parties (sometimes government or quasigovernment institutions). Therefore, we provide below some guidelines on how to learn from external incidents.

18.3.1 Do not react, do not panic, and do not implement half-baked ideas

In response to an incident we must make sure we understand reasonably well what had gone wrong before we start implementing fixes or solutions. Some often-heard rash suggestions in the aftermath of an incident in the oil and gas industry are:

- Put gas detectors everywhere.
- Install duplicate, independent level gauges in all our tanks.
- Provide water deluge protection on all our hot pumps.
- Double or triple the fire water system capacity in the refinery.
- Convert all buildings to blast-resistant design.
- Get rid of all portacabins.

A panic-stricken, rash reaction to an incident is not likely to generate long-term, sustainable risk reduction measures in your company. You must make sure that you have collected detailed information about the incident, you have analyzed the incident properly and then you have proposed mitigation measures that you believe are cost-effective and will thus benefit the company in the long run.

18.3.2 Form a task force or a small working group

It is highly recommended that a focused working group is created to study incidents. This is the most effective way of deriving the maximum benefit from external incidents. Select individuals from various disciplines. This will make sure that the details of the incident can be understood well by the group, jointly. Any recommendations that this task force/group generates will have a better chance of being implemented because of the acceptability by the departments that were a part of the task force.

18.3.3 Make one person responsible

Even in a task force situation, one person needs to be assigned the lead role. Also, in many cases the incident reports that need to be reviewed can be very bulky and it is perhaps not advisable to ask every member of the task force to completely read through every word in the report. Nonetheless, at least one person in the task force should go through the entire report quite carefully so that he/she can provide worthwhile advice to the task force (Table 18.1).

Similarly, it can be an equally daunting task to search for and read through information related to major incidents available on the internet. For example, there are numerous websites and books containing information and various points of views on the Bhopal tragedy alone [7]; and a great deal of time and resources will be needed to extract information pertinent to one's own industry and risk exposures.

TABLE 18.1 Major incident reports: size of reports.

No	Incident investigation report	Number of pages in the report
1	Piper Alpha (1988) by Lord Cullen	484 pages of A4 size and 30 pages of A3 size drawings and tables in two volumes
2	The CSB Report of the BP Texas City Incident (of March 23, 2005)	341
3	The James Baker Report of the BP Texas City Incident	374
4	The Esso Longford Incident in Victoria province of Australia	290
5	Buncefield Incident UK	116 (Vol 1) + 208 (Vol 2a) + 203 (Vol 2b)
6	Indian Oil Corporation Jaipur, Tank fire	195 + Appendices
7	BP Deepwater Horizon	192 + Numerous Appendices
8	Chevron Richmond Refinery (the CSB Investigation Reports)	70 (Interim Report) + 132 (Final Report) +126 (Regulatory Report)

18.3.4 View animation videos in a group instead of individually

Try to arrange joint viewings of any videos or PowerPoint presentations or animations of the incident being studies. The discussion that takes place after such viewings is extremely valuable in the learning process. It is through a common understanding of the details of the incident that the group will be able to generate meaningful and effective risk reduction measures.

18.4 How to make effective recommendations

Recommendations aimed at improving safety of plants and reliability of assets can come from a variety sources such as internal and external reviews and audits, management tours of sites, new standards and industry best practices, PSM requirements, insurance surveys and operational excellence (OE) reviews.

For safety and reliability enhancement it is essential that all recommendations are carefully reviewed to understand the primary intent, clearly identify the actions that are required to implement each recommendation in a timely and cost-efficient manner. During incident investigation training courses, it is emphasized to all attendees that the acceptability of a recommendation by the action party can be greatly enhanced if the recommendation is worded using the SMARTER rule.

SMARTER is an acronym that describes the essential features that need to be built into a recommendation to ensure that it is clearly understood by those who will be ultimately responsible for implementing it. The acronym stands for:

S	**Specific:** Who will do what and when?
M	**Measurable:** Can the corrective action be measured?
A	**Accountable:** Does the person assigned have resources and authority?
R	**Reasonable:** Is the corrective action practical?
T	**Timely:** Is the due date soon enough to prevent recurrence of problem?
E	**Effective:** Is it cost-effective per LCCA (Life-Cycle Cost Analysis)?
R	**Reviewed:** Has someone independent reviewed it prior to its approval?

18.5 Concluding remarks: lessons from history

As mentioned earlier, we can learn valuable lessons from incidents that have happened in the process industry (including oil and gas, petrochemicals, chemicals, and onshore and offshore installations). These incidents created a great deal of public interest at the time of their occurrence; some are etched forever in the memories of people affected by them, and all of them have gone into the history books of safety, loss prevention or allied subjects. In examining these, one should not delve unnecessarily in any controversial issues of culpability or any partisan agenda. The purpose must be to merely extract pertinent lessons from them in order to prevent or minimize chances of occurrence in your own organization.

The Feyzin LPG BLEVE incident resulting in 18 fatalities occurred back in 1966 in a French refinery where the operator was unable to close a drain valve due to valve freezing open. This incident triggered our pursuit for a greater understanding of the risks associated with bulk storage of LPG. The phenomenon called "BLEVE" had not really been well understood prior to Feyzin. There was a fundamental error in thinking that a relief valve on an LPG sphere, designed and sized for the "external fire" contingency, was capable of providing protection regardless of the overall fire exposure duration. It was not clearly recognized at the time that LPG would generate a ground-hugging pancake shaped vapor cloud that could drift over larger distances and find a source of ignition. The significant hazards associated with the sphere dewatering procedure were not recognized. It is a pity that even with the lessons drawn from Feyzin, vessel BLEVE's caused the heavy losses suffered in November 1984 at the LPG terminal belonging to PEMEX (the national oil company of Mexico)—over 500 people were killed in that accident!

The Flixborough incident forced the researchers into recognizing the reality of the "unconfined" vapor cloud explosion (UVCE). Prior to this incident, an UVCE was considered somewhat of a theoretical possibility—Flixborough changed that overnight. The UK Government had promulgated "The Health and Safety at Work Act" in that year (1974) and this incident brought home in a shocking way the need for employers to be proactive when dealing with the health and safety of their work forces. Another major lesson from this incident was that changes and modifications to plant and equipment, whether temporary or permanent, must be "managed" in a thorough and systematic manner.

The Seveso incident was the primary motivator in the EEC's adoption of the Seveso Directive that enshrines into law the public's right to know about the hazards posed by the plants and industrial installations in their midst.

Bhopal, and its aftermath, will remain a black mark in the history of our industry. Management that was so single-mindedly driven by short term profit maximization and cost-cutting with no regard to safety—how else could one justify the switching off the A/C unit of the tank to save a few rupees—has no business running our industry.

Piper Alpha taught us that adherence to the work permit system must always be ensured, without exception. The UK Government passed extensive legislation related to safety at offshore installations following the Piper Alpha disaster. The incident was very thoroughly investigated by a royal commission established for this purpose, headed by Lord Cullen. His report provides excellent details and a careful analysis of the root causes of this incident.

The Houston Phillips explosion acted as the wakeup call for the American Petroleum Institute who soon after the incident issued their recommended practice RP-750 (1990) detailing the PSM system that had been originally conceived and popularized by the Center for Chemical Process Safety of the American Institute of Chemical Engineers (ca 1985, immediately post-Bhopal). Not only the API, but the US government bodies had to respond also which they did in the form of Federal Register OSHA 1910.119 (PSM) in February 1992.

The BP Texas City incident highlights to us the need to eliminate all on-plot relieving vents in favor of properly designed flare systems. We must make sure that when we permit "oil-in" during the start-up of a unit after a turnaround and inspection phase there are no personnel present on site who are not directly involved in the unit start-up. Furthermore, we must make sure our senior management clearly understands the difference between "personnel safety" and "process safety."

Risks to process plants, especially pipelines, from vandals and saboteurs have grown manifold to become a serious cause for concern over the past decade. Transportation of crude and products via pipelines in many parts of the world was seen as the most cost-effective option—for example: lines from the Russian Federation to Europe, Nigeria, Angola, South Africa and neighboring countries, Southern Iraq, routing from Oman to India, routing from Iran to India, and a host of other proposals.

In general, the history of incidents teaches us that in the field of loss prevention in the process industry, there are a few key features related to layout and design, which tend to enhance the intrinsic safety of a plant. For example:

- Proper **spacing** (between equipment/units)
- Proper **size** (pipe/vessel size/wall thickness, etc.)
- Proper **steel** (correct metallurgy)

These features, when incorporated into the layout and design of a refinery or process plant, provide a significant degree of safety by mitigating the consequences of process deviations and other incidents. Furthermore, they are immune from the adverse effects of human error or other uncalled-for human intervention. In well laid-out refineries, risk exposures will be limited because of the generous inter-unit distances. The estimated maximum loss (EML) calculations carried out by the insurers in such cases reflect this lower risk, which, in turn, translates into lower premiums.

The intrinsic safety principle aimed at avoiding incidents needs to be fully supported by efficiently working management systems and robust leadership accountability philosophy in a company. Even though the number of incidents in modern times might be decreasing slightly, the cost per incident keeps on escalating. Likewise, a greater awareness and sharing of information with public is a must. It makes it far more important for company leadership to promote the concept of "Zero Accidents and Zero Incidents" as viable corporate targets. Proactive leadership does not wait for an incident to occur or the public to complain before striving for excellence in all its operations and business practices. In modern context, our license to operate depends on it.

In conclusion, and to quote Otto Bismarck, "Fools say that they learn from experience. I prefer to profit by others' experience."

Acknowledgement

The author has drawn freely from various publicly available websites in compiling the information contained in this paper for which he is grateful to the entities that own and maintain these sites.

References

[1] T. Kletz, Still Going Wrong!, Elsevier, 2003. Page vii.
[2] Chemical Safety Board (CSB) of USA, various incident investigation reports and explanatory videos and animations of incidents; CSB website www.csb.gov. Incident Report: LPG Fire at Valero – McKee Refinery dated February 16, 2007.
[3] Journal of Loss Prevention in the Process Industry, Volume 41, May 2016, Pages 186-193: Examination of fire and related accidents in Skikda Oil Refinery for the period 2002–2013 by Samia Chettouh, Rachida Hamzi, and Khemissi Benaroua.
[4] Llory, M. Acidentes industriais: o custo do silêncio. Rio de Janeiro: Multimais, 1999a. Gérard Mendel quoted in the Editorial Section.
[5] CSP (Cadernos de Saúde Pública) – Reports in Public Health, Open Publication Number Cad. Saúde Pública 2018; 34(4):e00034618: Revisiting the P-36 oil rig accident 15 years later: from management of incidental and accidental situations to organizational factors by Marcelo Gonçalves Figueiredo, Denise Alvarez, and Ricardo Nunes Adams.
[6] UK Health and Safety Executive (HSE) – Report - Offshore Injury, Ill Health and Incident Statistics 2004/2005. Data extracted from Table 2 on Page 18.
[7] Some websites related to the Bhopal tragedy:

Further readings

https://en.wikipedia.org/wiki/Bhopal_disaster.
Bhopal Information Union Carbide's pages.
Bhopal.net a counter view.
Bhopal.org another counter view.
Bhopal text, photographs, discussions and links.

Long-term recovery from the Bhopal disaster book chapter 1996.
The Incident at Bhopal online essay for the unscrupulous student.
https://en.wikipedia.org/wiki/Bhopal:_A_Prayer_for_Rain.
https://www.youtube.com/watch?v = HsuUQzhP2Ds.
https://www.theatlantic.com/photo/2014/12/bhopal-the-worlds-worst-industrial-disaster-30-years-later/100864/ all accessed in 1st September 2019.

Gauging the effectiveness of implementation and measuring the performance of PSM activities

Mark Hodgkinson

Bapco, Kingdom of Bahrian

19.1 Introduction

This is a chapter that draws on all the discussion in the previous chapters and will give an in-depth discussion on how we measure the effectiveness of our process safety management (PSM) system practically. This remains the biggest and most formidable challenge on how we prove that the PSM system is in place and delivering the kind of added value and effective loss prevention which was intended for it.

It also discusses the challenges faced in defining, monitoring and improving key performance indicators on human factors that play a key role in achieving and sustaining a mature and generative organizational process safety culture. This chapter will also explore how systems and system effectiveness declines over time and what strategies can be employed on how to keep the system effective.

Ensuring that PSM processes are designed to be human-error tolerant and are effectively deployed is the "management" part of PSM. Many organizations have deployed PSM processes but faced entropy in the process application with resultant incidents [1].

Several key issues in avoiding process safety events have been identified in the literature previously. CCPS state that these factors are the most relevant [1].

Process Safety Management and Human Factors.
DOI: https://doi.org/10.1016/B978-0-12-818109-6.00019-8

- An understanding of the risks associated with the facilities and operations.
- An understanding of the demand for process safety activities and the resources needed for these activities.
- An understanding of how process safety activities are influenced by the process safety culture within the organization.
- Focus on process safety effectiveness and efficiency. Use metrics to measure performance and efficiency so that finite resources can be applied in a prioritized manner.

This chapter focuses upon the last point around assessing process effectiveness and efficiency. Lenses of design, compliance (standardization), risk, and performance are used to build up the effectiveness landscape, while, the functionality of process tools, resources, roles and responsibilities play a part in process efficiency.

Efficiency and Effectiveness are linked in that humans will find a way to make an inefficient process into a noncomplaint process and eventually an unstandardized and faulty process (Table 19.1) (e.g., going from B to C in Table 19.1).

Therefore, both effectiveness and efficiency are important because they affect each other and if not understood and managed effectively can counter-effect one another.

19.2 PSM assurance

Line management has the primary responsibility to assure that its operation meets PSM requirements that are applicable to its processes, facilities or activities. The PSM Assurance process is the method by which leaders assure that necessary safeguards are in place and functioning. PSM assurance process tends to be multilayered with some levels of redundancy due to its criticality in managing high consequence risks. However, compliance assurance efforts should be risk based so they are efficient and effective. Striking the balance is the key (Fig. 19.1).

19.3 Design of PSM

While clearly, PSM process should be designed to meet given requirements, they should also be designed to have a special focus on being error tolerant. The PSM Assurance Process should firstly map all the PSM processes and check that the following design aspects are met. If not, the process should be improved. In PSM, there are usually accessible best practice templates for the processes that can be used as a

TABLE 19.1 The relationship between effectiveness and efficacy in processes.

	Efficient (suitability, adequacy),	Not efficient
Effective (Authorized, documented, Measured, Accountabilities clear)	A: Aim is to design and implement systems that lie in this quadrant	B: Process meets requirement but has unnecessary restrictions, constraints to the business, Poor functionality in tools or perhaps roles and responsibilities not clear.
Not Effective	C: People "cutting corners" to get work done	D: Poorly designed processes poorly implemented.

FIGURE 19.1 The PSM assurance process pyramid.

starting point. See validation section and refer to Table 19.2 of design for a practical guide to checking the design.

19.4 Supervision of PSM

Supervision in this sense, refers to first line supervision of work. Many companies have an Operational Excellence Leadership Model or

TABLE 19.2 Effectiveness evaluation criteria (exhaustive).

	Less than satisfactory	Satisfactory	Good
Scope, purpose and objectives	None or a few of the objectives are being met, or the organization is not tracking and does not know if the Process objectives are being met.	PSM Process objectives are met.	Most or all the Process objectives are consistently met and performance trends are in the desired direction.
Procedures	The day to day work of the Process is not consistent with the design. The design may be incomplete, represent an ideal situation, etc., but does not reflect "what actually goes on in the field".	Daily work activities (procedures) are consistent with design and are consistently followed; some gaps exist.	Daily work activities (procedures) are consistent with the design; only minor deviations occur.
	Employees erratically follow procedures or procedures are only partially consistent with design; includes many gaps.	Risk evaluations are conducted to prioritize activities.	Appropriate emphasis is placed on identifying higher risk activities.
	Risk evaluations are not conducted to prioritize activities.		
Resources, roles and responsibilities	Personnel are unaware of their roles or have incorrectly assigned roles.	Personnel are aware of their roles.	Personnel are aware of their roles and take an active interest in their position.
	Not all personnel assigned have appropriate skills.	Personnel have appropriate skills for required tasks but may not be fully utilized.	Personnel have appropriate skills for the required tasks.
	Resources to improve the design or effectiveness are either inadequate or lacking appropriate priority.	Resources to improve effectiveness are in place.	Resources to improve the design or effectiveness are adequate for the risk. There is a pattern of adequate resources through time with

(Continued)

TABLE 19.2   (Continued)

	Less than satisfactory	Satisfactory	Good
			only minor periods of time being under resourced.
	Emphasis is placed on what is "easy" to do, not necessarily what will provide the most impact relative to the risk.		Appropriate emphasis is placed on higher risk activities; priorities are consistent with other tasks and activities.
Measurement and verification	• Minimal or informal performance measurements and verification occur for the Processes. Metrics do not address verification (i.e., lagging indicators are not developed).	Ongoing performance measurements and verification occur as documented in the design. Frequency may vary from process to process.	• Metrics (both leading and lagging indicators) are recorded and tracked as documented in the design. Frequency is appropriate and the data analysis creates actions to keep the Process on track to achieve desired results.
Continual improvement	• Few, if any, PSM design reviews are conducted to identify performance gaps. If there are design reviews, they are informal or not documented. • Improvements are not identified, or if they are, then action is not taken. • Tracking for continual improvement is limited and the backlog of actions indicates little, if any, improvements	PSM design reviews are conducted for continual improvement. Gaps are identified, tracked and closed.	• PSM design reviews occur regularly as documented in the design. • Improvements to Processes are identified and tracked to ensure implementation for continual improvement. The backlog of incomplete but identified improvements is consistent with the overall PSM Performance and associated risks.

(Continued)

TABLE 19.2 (Continued)

	Less than satisfactory	Satisfactory	Good
	in a Process's design or effectiveness have occurred. Action is taken only when external Reviews occur.		There is a pattern of identifying, implementing and closing out actions in a timely manner.

PSM Leadership Model that sets expectations of leaders including supervisors (Table 19.3).

The guideline minimum level of direct contact supervision is about 30% of a front-line supervisor's time per day. The emphasis of this supervision should be ensuring that all employees are always doing the right thing the right way. Supervision of PSM processes is a key control.

A practical guideline for people in supervisor roles reading this chapter and wishing to deliver PSM expectations is outlined below (Table 19.4).

Supervisors should seek to standardize the normal days' work and eliminate variation. In this way, the workforce can more easily identify abnormal situations and act upon them.

19.5 Verification of PSM

Verification refers to checking that the right work is being done the right way. Generally, it refers to the duties of Superintendents and Managers that ensure that (a) Supervision is being carried out by first line supervisors to standard, (b) that operational discipline is being practiced. Some organization refer to this as "Active Monitoring", Operational Audits or in Six Sigma a "Gamba Walk".

Verifications that one's own team is on process is normally called a "Check." Verifications of other teams is normally called an "Inspection" (e.g., Carried out by the PSM team). Both checks and inspections are critical as any significant difference between the average results of checks and inspections could indicate a finding in verifier capability. Various bodies such as UK HSE have good tools to use as a starting point.

The expectations of leaders with respect to verification should be risk based depending on organization role but also their expertise. Targeted

TABLE 19.3 Typical PSM leadership expectations [2] (see Chevron OE Expectations).

PSM expectations mostly applicable to executive leadership	PSM expectations also applicable to supervisor leadership
• Executives and managers establish a PSM culture. • PSM assurance efforts are in place and effective (the subject of this chapter). • Executives and Managers personally direct continual improvement in both PSM process designed (to be error tolerant) and effectiveness. • Leaders establish and execute a plan for engaging the workforce and raising PSM vigilance.	• Leaders remove barriers to performance • Leaders demonstrate knowledge of PSM • Leaders define and communicate expectations for assurance of compliance with PSM requirements • All leaders develop and assure the competency of the workforce • All leaders provide accountability for complying with PSM requirements • Participate in PSM Assurance Activities.

direct positive feedback to employees works the best provided the verifier has the skill to truly identify good work. It is the recommended methodology for performing Verifications rather than looking for faults and errors.

Experience has shown that, for successful verification activities to be achieved they must be standardized and tracked. Informal verification practices normally fail over time. Thus, when applying the systemized and tracked verifications we must ensure;

• The experience and expertise of verifiers.
• The results of the verifications can be analyzed.
• The workforce understands the verification activity is part of the process.
• The "Process Custodian" has good quality diagnostics to make decisions upon.

   Leaders are expected to be trained (as shown in compliance records) and fully understand each process so that they know what "good" looks like.

   Good PSM examples of verification activities are checks and Inspections are;

• Permit to work Checklist
• Lock Out Tag Out Checklist
• MOC Review Checklist
• Management Review of Incidents after all actions have been taken.

TABLE 19.4 Practical PSM supervision.

Supervisor PSM activity	Guidelines
Leaders remove barriers to performance	Check if your team has access to all the Process Safety Information they require. If not, rectify the situation or apply Stop Work Authority.
Leaders demonstrate knowledge of PSM	Leverage any incident material supplied by your PSM team. Talk about the root causes of the incident with your team in toolbox talks.
Leaders define and communicate expectations for assurance of compliance with PSM requirements. All leaders provide accountability for complying with PSM requirements	Make it very clear to your team that doing the right thing the right way is what you expect.
	Communicate key messages like "MOC not in-kind changes" and energy isolation steps. Keep communicating as is surprising how little people listen. Focus on positive reinforcement of people doing the right thing. If there has been an error or an incident, call for help from the PSM/Safety/OE groups to see if human factors were involved. Avoid jumping to conclusions or blaming people for errors (there might be other factors). Avoid hiding the error as fixing the root cause could save a life.
All leaders develop and assure the competency of the workforce	At a minimum ensure that all compliance training is done by your team. Ask open ended questions to determine if your people understand PSM processes that they work within.
Participate in PSM Assurance Activities	See other sections of this chapter.

19.6 Metrics for PSM

There are now several very good publications on PSM metrics [see CCPS, API 754]. We have found that the key issue is striking the balance between leading and lagging metrics. While, it is common to be subject to criticism of having too many metrics, PSM processes are diverse and complex. Hopkins [3] made the point after the Texas City Process Safety Event that leading indicators (e.g., the compliance metric: Number of PTW Verification) and lagging indicators (e.g., The performance metric: Incidents where PTW failure was shown as a root cause) usually do not give enough forewarning. The results of the verifications (termed "Risk" metrics) are the best predictor of the performance. Hence, by the time we have leading, risk and lagging metrics we can have quite a number. Only a sub-set should be shown to reviewed by top management, but the process custodians should be fully aware of the full picture.

We have found that having a proper KPI software with workflow for data providers to enter data and comments on the data, process custodian comments and displays that allow multiple year analysis and projections is much better that human operated Excel sheets.

Here is an example of such a system (Figs. 19.2–19.4). Examples of Metrics commonly used is shown in Table 19.5.

19.7 Audits of PSM

An Audit is a systematic, independent, documented and objective review to evaluate its current status relative to the defined audit criteria. These requirements include using trained, competent auditors, use of a protocols or checklist built upon standard expectations, documenting

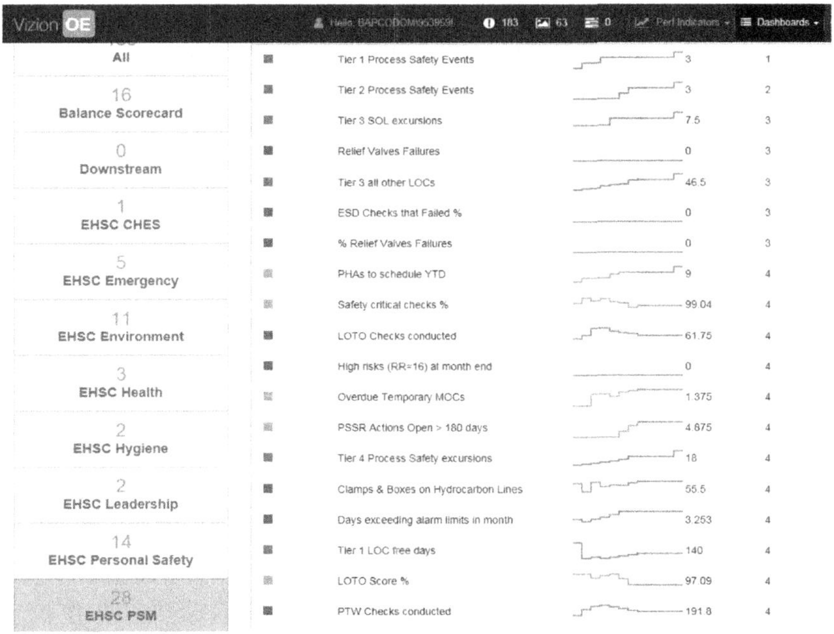

FIGURE 19.2 Explorer view of a refineries PSM metrics.

FIGURE 19.3   Example of functionality to show multiple years and projections.

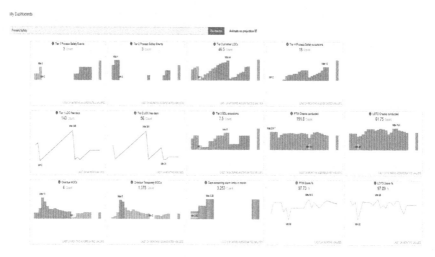

FIGURE 19.4   Example of higher level dashboards for general work force (with drill down).

and reporting results to leadership, and tracking corrective actions to closure.

While it is best practice to incorporate the PSM Audits into the sites overall management system internal audits, one must be careful not to oversimplify the audit of Process Safety application. The British 5 Star Process Safety Audit protocol is one type of holistic audit that gives a score that can be optimized. However, a sampling exercise where the whole data set of a high-risk unit (highest inventory of volatiles, highest

TABLE 19.5   Examples of PSM metrics.

PSM process	Leading	Health/risk	Lagging
Process safety information	Drawings and data sheet update conformance		Incidents where a root cause includes PSI found to be missing or wrong
Process hazard analysis	PHA to schedule Open PHA actions Overdue PHA actions	PHA Actions > Risk Rank X	Incident investigations that show the root cause includes PHA was not done or action item from PHA was not implemented in the specified time period
Operating procedures	Ops and Mtce Document Reviews Over due	Ops and Mtce Documents changed per year	Incident investigations that show the root cause includes poor or missing document
Training	Mandatory Training conformance		Incidents where a root cause includes Training Issues
Contractors safety management	% Contractor Safety Meetings to schedule	Number of safety violations by contractor	Contractors Statistics (LTI, NLTI, FA)
Pre-start up safety review	PSSR Actions overdue (190 days)		Lagging: PSSR faults -Post Start up PSSR incidents captured under Tier 4: PSM incidents
Mechanical/ asset integrity	Compliance of Statutory Checks /Inspections on equipment and safety critical devices LOTO, PTW checks	The result of the checks Safe Operating Limit Excursions Boxes/Clamps on H/C lines Unexpected Corrosion events Alarms levels	Loss of Containments T1,T2, T3
Safe work practice PTW, isolation, confined	Leading: Conformance to standard for working at heights, Excavation and Falling objects. BBS	% Behavioral showing PPE Compliance PTW, LOTO, Safety Audit/ checks/	Lagging: Injury Frequency Rate

(Continued)

TABLE 19.5 (Continued)

PSM process	Leading	Health/risk	Lagging
space, stop work, excavation. others		Inspection findings	
Management of change	MOC's overdue Temporary MOC's O/D	MOC Check findings (perhaps changes that did not use MOC)	Incidents where MOC failure was a Root Cause
Incident investigation	Overdue Incidents Investigations Near Miss to Incident ratio	Incident Investigation Action close out rate	Repeat Incidents
Emergency planning and response	Emergency Exercises Fire Drills	Fire Equipment Readiness	Fires API 754 T1, T2
Compliance audit	Audits to schedule	Audit findings	Tier 1 and 2 HSE Incidents
Governance and audit		CAPA close out rate	

pressure) is sampled is another approach. In this case one then looks at the PSI details for that unit, the MOC history, the incident history, the T&I and PSSR history, MI history, relief system design, HAZOP history, operating procedures, training materials and an extensive site inspection. Clearly, engineering skills are required for the latter in-depth audit.

External audits often by Insurance underwriters, safety case officials or legal bodies will normally be staffed by subject matter experts. Hence, it is essential to ensure PSM expertise on internal audits as well.

19.8 Verification workshops

A new leading practice to assure that controls are in pace and functioning is utilize workshops with the employees working in the process. One methodology that we have had success with is the highly visual Bow Tie process. With Bow Ties, you can target top risks (Major Accident Hazards) so the workshop focuses on controls, control effectiveness and barriers. A qualified Process SME conducts the Safeguard Assurance workshops.

The workshop is performed by gathering input from subject matter experts and the actual people performing the work and/or controls. The team then analyzes which safeguards and verifications should be in place to prevent or mitigate a Hazard. Unstandardized or missing process steps can often be discovered.

Document the results of the workshops with respect to the design and effectiveness of controls. Implement the actions resulting from the workshops. The results of a Bow Tie workshop can be also used for a wider PSM alignment and awareness.

19.9 Validation

Validation is usually conducted by the process Custodian. It is the process where we "Validate" that the process intelligence is reliable and telling a true story. It is usually an annual task and involves following a written protocol that looks at all the aspects talked about above. It is a form of Self-Assessment at an operational level. It always involves interview with a crosscut of the organization together with diagnostics on the checks, inspections, audits, incidents and metrics. A matrix is typically used to provide an overview of the findings as overall, design and effectiveness (Table 19.6).

19.10 Management review meetings

Management Review Meetings (for management systems) are a form of governance. The management review process requires top management to periodically review PSM to ensure its continuing suitability, adequacy, effectiveness and alignment with the strategic direction of the organization. This was originally lacking in the PSM design but has been recognized as being required [1].

Management Review meetings:

- Determine and evaluate PSM performance
- Determine the need for change and improvement
- Determine the suitability of the policies and the objectives
    Some guidelines are;
- An Operation Intelligent Tool (see below) can facilitate effectiveness and efficiency of information analysis.
- Governance meetings are not problem solving meetings (e.g., Bad Actors) and not planning meetings (e.g., Priority Output). The must be focused upon Assessing Operational Excellence of PSM Processes with respect their suitability of purpose, adequacy, compliance,

TABLE 19.6 Design evaluation criteria (non-exhaustive).

	Less than satisfactory	Satisfactory	Good
Scope, purpose and objectives	Process contains no scope, purpose or objective, or it is unclear, poorly designed and incomplete. There is no indication of what's included and what's excluded, or it is unclear. Links to other Processes are not identified. Specific, measurable objectives are not stated.	The scope, purpose and objectives are adequate, however, some of aspects (e.g., links to other Processes, specific objectives, what's in/what's out) still need to be defined or are inappropriate.	All aspects of the scope, purpose and objectives are completely defined and appropriate. There is confidence that work will not "fall through the cracks". There is clarity of what's included and what's excluded and how this Process is linked to others. The purpose of the Process is clearly defined with a statement of intent and specific measurable objectives.
Procedures	Few procedures are documented and there is a reliance on "institutional knowledge". Many do not have sufficient design details to work when implemented. There is no central location pointing to existing procedures and there is no sense of "workflow". There is little or no consideration of the risks of the work, the skills of the employees and other relevant factors in the design.	Highest priority procedures (highest risk) are in place with sufficient design details to work when implemented. Additional procedures need to be completed. The workflow may still need to be developed for the Process. Design of procedures have considered the risk of the work.	Procedures are in place and documented with sufficient design details to work when implemented. The workflow of the Process is in place and easily followed. The rigor of the design is appropriate for the level of risk: The higher the risk or more non-routine the task, the more rigor is required of the design.

(Continued)

TABLE 19.6 (Continued)

	Less than satisfactory	Satisfactory	Good
Resources, roles and responsibilities	Few roles and responsibilities have been defined for this Process. Those responsible may not know it, or it is a low priority add-on assignment. The Process has no resources assigned.	Highest priority roles and responsibilities for PSM have been defined. Additional definition is required.	Roles and responsibilities for the Process have been clearly defined. There is a distinction between those who are responsible for the design (maintain, administer and define who and how) and those responsible for performance (conduct/carry out the tasks). The Process is fully resourced: Those assigned know it and have appropriated priorities to make desired progress.
Measurement and verification	Few or no objectives/ results measures (lagging indicators) are identified.	Some metrics are in place to track performance, typically the results measures (lagging indicators).	Both leading (measurement) and lagging (verification) indicators are identified and appropriate.
	Those identified are inappropriate for the Process. The critical components of the Process (leading indicators) are not measured or are inappropriate. In other words, there are no checks to see that the process is operating effectively.	Other metrics need to be developed, typically the leading indicators (determine that the critical components of the Process are adequate).	They are measured, tracked and evaluated. Leading indicators have been developed to measure the critical components of the Process. In other words, there is a check to ensure that the process is operating effectively.
Continual improvement	Process design reviews are not conducted or, if they have, they are informal or not documented.	PSM design reviews have been conducted and actions are under way to implement continual improvement.	There is clarity as to how the Process is periodically and formally reviewed and evaluated.

(Continued)

TABLE 19.6 (Continued)

	Less than satisfactory	Satisfactory	Good
			Documentation of periodic PSM design reviews exists.
	Improvement steps are not identified; or, if they are, then they are not implemented (no actions taken) or tracked to completion.		Steps to improve processes have been identified, implemented and tracked to completion for continual improvement.

effectiveness and efficiency and then to direct continuous improvement.

- They should be held at least annually.
- Actions must be held on the business record.

The purpose and outcome of the management review should be continual improvement of the PSM.

19.11 Operational intelligence

Operational intelligence (OI) is a category of real-time dynamic, business analytics that delivers visibility and insight into data, streaming events and business operations. It is ideally suited to monitoring the performance of PSM activities. It differs from business intelligence (BI) which is about analyzing what has happened to date and looking for efficiencies to optimize the business in the future.

OI information feeds from PSM processes is somewhat dependent on automation of PSM tools which should bring efficiencies. We know from experience that high workloads on SME's to prepare for validation or management review activities is resisted and often fails. Management of process safety needs to me made efficient if it is to remain effective.

19.12 Closing thoughts

In some of the progressive organizations around the world, PSM has been integrated with operational excellence. Operational excellence is fundamental in trying to ensure PSM effectiveness. In this chapter, effectiveness has been discussed from an assurance, design, supervision and leadership as well as metric monitoring and governance assurance perspectives. All are important to consider when developing PSM and Operational excellence and integrity management systems, which are designed to ultimately deliver value and reliability to an organization making it safer over time.

References

[1] CCPS Risk Based Process Safety (RBPS).
[2] Chevron OE Expectations: Personal communication.
[3] https://static1.squarespace.com/static/536316c1e4b03715f2388919/t/5602010ae4b0-ceaec9081a62/1442971914428/WP + 53 + Thinking + About + Process + Safety + Indicators.pdf.

Human errors, organization culture, and leadership

Hirak Dutta

Formerly with Indian Oil Corporation Limited; Oil Industry Safety Directorate. (OSID); Ministry of Petroleum & NG; Nayara Energy Limited; Indian Society of HSE Porfessionals, Uttar Pradesh, India

20.1 Introduction

In my last four decades of association with industry, and more specifically with the oil and gas industry, I have witnessed some major catastrophic failures. Luckily, I had the opportunity of participating in several investigations of some of major incidents. Some of the incident investigations were carried out by me as team leader and some others as a member of the investigation team. I had overseen implementation of path-breaking recommendations in my role as Head of Oil Industry Safety Directorate in India and the improvements that have been brought about in marketing terminals.

As I reflect and take a closer look at the root cause of the incidents, which I have described in Chapter 8, Asset and mechanical integrity management, besides others, human failures/human errors also contributed to unsafe situations. The question that crops-up upper most to mind is why do humans commit errors? Is it due to the competency gap? Is it because of an organization's culture? Is there any evidence of human error vis-à-vis organizational culture?

Process Safety Management and Human Factors.
DOI: https://doi.org/10.1016/B978-0-12-818109-6.00020-4

269

20.2 Human error and organizational culture

While the other authors in this book have highlighted and in greater detail, I have tried to classify the human error in three broad categories, viz., action/checking errors, diagnostic errors and decision errors. These classifications are based on many incidents that happened around the world. Action or Checking errors happen due to not taking appropriate actions or wrong actions and/or omitting some checks. The checks that are very important to ensure compliance to the systems and procedures; checks are paramount in ensuring standard operating procedures (SOPs) are not disregarded or disrespected. Checks, however trivial, keep the system well-oiled. Diagnostic errors take place due to misinterpretation of data and thereby judgmental mistakes in taking appropriate corrective actions.

Decision errors on the other hand epitomize wrong decisions. Wrong decision-making or not taking timely decisions adversely impacts operations. And an industry dealing with highly flammable products and complex operations any delay in taking correct decision may result in devastating consequences.

Let me cite an example. Cracking reactions are exothermic in nature. In refining or petrochemical units say in a hydrocracker unit (HCU) or DHDT unit: runaway reactions lead to severe operational upsets so much so that if uncontrolled it could end up in a major disaster. The temperature rise could be so uncontrolled and rapid that ultimately the thick steel metallurgy of the reactor may fail leading to escape of hydrogen to the atmosphere. This release of hydrogen and hydrocarbon vapors at such high pressures and temperatures could lead to a major catastrophic event in the plant. HCU normally operates at 120−150 bars at a temperature of 350°C−400°C in the presence of the highly flammable and explosive hydrogen gas environment.

To both control and mitigate such severe conditions viz. runaway reactions in the HCU unit, several layers of protection systems are envisaged in design. The steps include increase the hydrogen quench in the reactor beds; cut off the furnace at the upstream of the HCU reactor; withdraw the feed and/or depressurize the system to flare etc. Hydrogen quench acts a coolant to the reactor bed and increasing the flow of hydrogen quench in the catalyst bed is aimed at cooling the reactor thus preventing abnormal rise in catalyst bed temperature. In case the above step does not arrest rapid increase in catalyst bed temperature, one has to cut off almost immediately the furnace upstream of the reactor to cut off any additional heat input to the reactor.

The next step to arrest the imminent damage is to withdraw the feed which will tantamount to shut down the unit. The ultimate choice a supervisor makes is releasing the reactor pressure which is technically called dumping the reactor pressure to flare header of the refinery. Normally two dump valves are provided to release the pressure at 21 kg/cm^2—the first dump valve releases pressure at 7 kg/cm^2 and the second at 14 kg/cm^2—until the reactor is completely depressurized. Depressurization ensures drastic reduction in hydrogen partial pressure thereby confirming exothermic reaction or runaway reaction is completely stopped. However, delay in taking actions could cause severe damage to the unit including a major catastrophe.

These are some of the design provisions to manage runaway reactions and supervisors in HCU unit are given training to handle such emergencies before they are given formal charge of operations. Sounds reasonably simple, but when such situation arises, from the author's experience, it is an extreme emergency situation.

The dilemmas before a supervisor or a plant manager are many. So, let us try to understand the dilemmas: a sincere supervisor/plant manager tries to avert plant shutdown. Shutting down HCU and restart-up is time consuming and profound production loss. At the back of his mind, the physic mindset is not to shut down the plant which inhibits him to take some of the drastic steps viz. withdrawing the feed and depressurizing reactor. Of course, coupled with this, the supervisor must also have prerequisite traits like maintaining cool and necessary expertise to tide over the situation and obviate a disaster.

The diagnostic error plays a paramount role. A supervisor/manager delaying the process of withdrawing feed end up in damaging the catalyst and in extreme conditions damaging the reactor itself. His or her ability to judiciously act which it is strongly believed has a direct bearing on his ability and more importantly on organization culture constantly buzzes his decision-making ability.

The relevant questions therefore are: Why do human makes error? Is human error due to:

1. Absence of safety leadership?
2. Lack of reporting system?
3. Too much high stress?
4. Lack of training and development?
5. Lack of motivation?
6. Lack of safety culture?

Safety culture of an organization could be driven by external and internal motivation. External motivation is driven by demands to meet the compliances, statutory rules/regulations and protocols, whereas internal motivation is about fulfilling the commitment: the commitment to its stakeholders, to its external and internal customers. The subtle difference is I follow the rule because I must follow, vis-à-vis, I follow the rules because I like to follow. And this makes a serious difference in terms of setting up and promulgating a just culture in the organization.

Let us focus on the culture of the organization in context with the above incident. Does an organization care for the training needs of a supervisor and honing the skills of its employees? Or the organization is interested in meeting the number of training man-days imparted to its employees. Are the managers given sufficient quality training opportunity? For example, are they given charge of simulator-based training before they are given charge of managing critical operations/units? Do the leaders in an organization care for and encourage candid and free discussions or rely on giving instructions to its employees? Do leaders in the organization walk the talk? Do they set examples in themselves for others to follow?

Leaders quite often in meetings mention shutting down the plant if unsafe conditions persist, but in private pull up the supervisors and plant manager for actually shutting down a running plant. We all have experienced enough of such leaders in organizations. They say something in open and go back on their words in private. Such leaders are not role models and certainly not inspirational leaders. Those leaders fail to enthuse their subordinates and fail to bring out the best of their colleagues and juniors.

Do leaders have trust and confidence in their teams? Or they operate from a higher platform, "I am simply better than you," mentality. A leader must have trust and confidence in himself and trust and confidence in his/her team. Such leaders are not insecure. They walk the talk.

The role of a leader is not merely to ask for results to meet the goals but a true leader provides adequate resources and directions to meet the targets. Leaders have a vison for the organization. They bring in systems and procedures and ensure they are implemented. They appreciate the good work being done. They provide challenging assignments and truly believe in teamwork: They follow therefore an Institutional approach rather than Individualistic approach. Such leaders inspire the team and meet the long-term superordinate goals of the organization (Fig. 20.1).

Management commitment is an essential ingredient for building a healthy safety culture in the organization. To achieve this objective, management must provide adequate resources coupled with unstinted support and provide direction. Any mismatch between what we speak and what we do may result in mistrust among employees and an environment not conducive to spontaneity. Though it is said that leadership should be visible, in fact leadership commitment must be so deep that it almost becomes invisible. Employees should not have any doubt on management's sincerity and genuine commitment that gets reflected in actions.

Safety culture deals with attitude and behavior of employees. Cultural change processes are slow by nature and involve coherent, consistent and

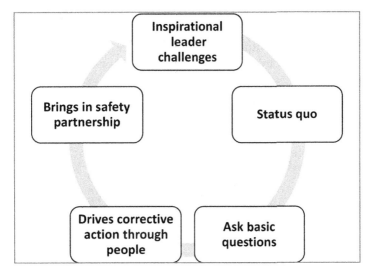

FIGURE 20.1   Inspirational leadership style.

sincere efforts. Promotion of safety culture is a long-term strategy which requires sustained effort and interest. Unless efforts are properly resourced, plans are prioritized and there is clarity on expectations from employees and contractors, chances of success will be limited.

Belief is deep seated in our mind. It is there in our system since time immemorial. It is not dependent of knowledge, proof or rationale. Mind simply accepts *belief* as truth because of past conditioning. When knowledge confronts belief, we behave from the standpoint of belief and not knowledge. It is easier said than believe that all accidents are preventable. Changing belief or introducing new belief will require tremendous amount of effort. Creating a safety culture requires management's total commitment, openness, transparency and sincerity. Effective communication, frequent conversation with the employees, friendliness and investing management's *time* will develop a healthy relationship which will keep focus and maintain awareness on safety issues and sustain high standard of safety culture.

20.3 The human paradoxes leading to incidents

Let us come back to the case of HCU runaway reaction. The supervisor does all sincere efforts to overcome the imminent emergency including avoiding shutting down of the unit. He sweats; he perspires; and he brings all his past experience, yet at times end up in a major accident. We must analyze as to why he behaves in such a way? What makes him procrastinate in decision-making in such a vital unit? The answer is loud and clear it is the culture of the organization that impairs the supervisor. The leadership style is not inspiring.

At this stage let me introduce some of the human paradoxes. Here I have tried to classify some of the paradoxes that is overtly visible in the industry:

- greed alibi optimization,
- deep-rooted human propensities,
- acquisition versus assimilation,
- averse to change management, and
- communication barriers.

Greed alibi optimization: Too much concern for throughput inertia thus the "production first syndrome." Such an attitude (within the organization) quite often proved to be counterproductive. Production first syndrome prevents the organization to ask the basic question: Production at what cost?

Deep-rooted human propensities: We refuse to learn though we appear to give lot of importance to the lessons learned. Our *know-all* attitude prevents us from learning and we do not learn from mistakes.

Acquisition versus assimilation: Acquiring state-of-the-art technology is easy if we have capital. And I believe oil companies are cash rich. But do we focus on assimilation of the state-of-the-art technology? Do we provide enough training to our employees to muster the technology? Assimilation of process and technology is paramount.

Averse to change management: Human nature is to resist change though we appear to appreciate only thing constant is change. Do we update drawings, manuals and documents: the changes that have been incorporated in our facilities? We are poor in documentation.

Communication barrier: Do we encourage freewheeling of discussion in our organization? Do we ask questions? Do we accept dissident views? Do we encourage joint problem-solving sessions? Managers are comfortable in giving instructions, and in the process, create a barrier in the team.

The mantra for success in today's complex business environment is "Listen to the Plant" and "Listen to the People." It is not a choice but a necessity. Gone are the days of giving instructions. Monologues discourage participation. Absence of participation leads to working in silos.

20.4 Closing thoughts

I conclude with this thought: In today's cutthroat competitive and complex business environment, culture of an organization and its leadership style plays a very vivacious role like never before. The onus is on creating a trusting and just culture. A culture that encourages joint problem-solving, a win-win proposition, and a culture where mistakes are accepted openly without the fear of punishment.

If we can create such an environment where all employees know the company's true safety commitment viz. safety policy, core values, and goals; where feedback is provided as a means to learning from mistakes, human beings are valued, and their contributions are celebrated: the organizations will surely excel.

Epilogue

This book provides a wide spectrum of facts, views, and experiential knowledge from a large group of diverse practitioners and academics. It has been a challenge as an editor to provide a review to bring many improvements in the text of the chapters as they were presented to me for my review and edits over more than a year. Not only are all the authors/contributors generally eloquent in their writing, but many of the thoughts they brought were very meaningful and useful.

Therefore, we tried a strategy of reading the chapters more as a reader rather than an editor at first. We wanted to ensure the impact from the experience in reading the text was to be eventually received by the reader. But in this closing chapter we provide a short synopsis of each chapter and explain the implications to the practitioner.

Chapter 1 provides a short introduction to the development of the PSM systems development with a historical overview. The chapter lists the main and core 14 elements of the PSM system as per the OSHA PSM standard. The chapter also tries to give some insights into the historical background on the incidents in the industry, which brought about an appreciation of developing a purposeful management system to manage process safety risks.

In the second chapter Dr. Ian Randle provides a very comprehensive introduction and provides the link between PSM and human factors. The chapter provides a very good introduction to the subject of human factors and the terminologies mainly used currently in industry. This chapter also provides good insights on why human factors are important to understand and consider in the context of safety. The management of human failures and an explanation to the modes of human failure and an explanation to the differences between errors and violations, which is quite fundamental to understanding the HF science. He also provided a very good explanation of the safety critical tasks and the analysis process that relate to the same.

Leadership is addressed in Chapter 3 and discussed the critical role of leaders in the context of PSM. The chapter gave three examples of PSM elements and the role that leadership must play in the effective implementation of the elements. How management can influence human behavior and human factors was also addressed. It must be appreciated that the role of leadership in organizations links with PSM leadership and governance within the context of the organization's strategy.

In Chapter 4, and purposefully placed in the early part of this experiential knowledge book is the *Awareness of Risk and the Normalization of Deviance*. Prof. Ron McLeod provided an excellent explanation to the ways and the underpinning reasoning that the deviation from what is clearly a safe—tried, tested, and approved—set of practices in process and behavioral safety. His chapter explains complacency comprehensively, including organizational, automation-induced as well as situational complacency. He goes on to explain highlighting with rational case studies the relationship between complacency and the normalization of deviation. This is a very significant chapter towards understanding human factors in action.

Moving on, but remaining on the subject of human factors, Jannette Edmonds provides an equally fascinating and interesting Chapter 5 focusing on *Competence Assurance and Organizational Learning*. In this chapter Jannette explains effectively why assuring human performance is a fundamental component of effective process safety assurance. This chapter is inoculated with various case study examples and provides for the practitioner very fundamental definitions and a very good understanding of the *Competence Development Continuum*. The importance and steps and process behind the development of a competency management system (CMS) and the implications of organizational learning are also addressed succinctly in this chapter.

In Chapter 6, Hari Kumar, a very experienced practitioner who has been involved in projects, operations, and development of corporate systems and standards for more than 35 years, addressed the *Integration of Human Factors in Hazard identification and Risk Assessment*. Risk assessments and management are fundamental skills for EHS and PSM practitioners. In this chapter, Hari presents the typical HSE Risk Assessment/ Studies in any major project life-cycle. In process safety HF in design, operational stages and human interfaces are explained.

The chapter reviews various aspects that practitioners, especially those who are advisors or managers, can benefit greatly from in terms of the management of process safety considering the human elements.

Then in Chapter 7, the pragmatic academicians Associate Prof. Dr. Risza Rusli and Dr. Mardhati Zainal provide a comprehensive text that looks at the *Inherent Safety Impact in Complying Process Safety Regulations and Reducing Human Error*. Starting with a discussion of the reasons for chemical industry safety incidents and addressing the need and significance of inherently safer design (ISD), the chapter goes on to explain how the reduction of human error through inherent safety can be undertaken. It gives a detailed process case study to illustrate the practicality of ISD. They proceeded with explaining that the practitioners understanding that human factors are involved at the very early stage of design and is therefore critical to eliminating or reducing potential catastrophic incidents. They conclude with that various ISD-led methodologies and the reduction in the inherent risk can have not only good risk reduction; but can also ensure better regulatory compliance as well as cost effective designs, etc. This is particularly important when looking at performance-based regulations.

Chapter 8 by Hirak Dutta explores asset integrity management. This chapter provides the foundations of asset integrity management, the prevalent models and discusses the People-Process-Technology Alignment to Achieve Process Safety Excellence. The chapter talks about the various factors that impact on integrity management and it concludes with a set of four major case studies and links the failing asset integrity related aspects.

In Chapter 9, Ram Goyal who is a highly experienced and seasoned expert in this field presents a very well structured and significantly important chapter on the management of change (MOC). The chapter explains with historical background case studies, types of MOC processes from cradle to grave, and discusses the management of organizational change (MoOC) which many technical reviews and literature on MOC fails, more often than not, to address adequately. The chapter has an excellent balance of case studies and explanations and concludes with a thought-provoking discussion around modern trends in risk tolerance.

Chapter 10 addresses the *Management of Risk Through Safe Work Practices* by Dr. Lutchman and Dr. Akula. This chapter goes through HF and risk management, behavioral safety, and managing human performance gaps but mainly focuses on risk management through safe work practices where the safe system of work, operating procedures, and safe work practices are developed effectively. This chapter guides practitioners to the establishment of safe work systems and helps both corporate and field-based staff in understanding the management system documentation. One relatively unique proposition this chapter provides is a focus on the effective communication of safe work practices and procedures.

In Chapter 11, a subject less understood by most practitioners is *Process Safety Information (PSI), Hazard Control and Communication*, presented comprehensively by the two academicians Prof. Azmi Mohd Shariff and Dr. Muhammad Athar. One unique and interesting thing this chapter presents to practitioners is a comparison between the pillars for process safety management and the PSM elements of four different and widely used process safety management systems. The PSI element is explained comprehensively and the implementation framework for PSI is also addressed. In the later part of the chapter the authors provide a very good overview of the aspect of human errors, which are applicable to PSI.

Then we move to Chapter 12, which addresses the *Pre-Start Up and Shutdown Safety Reviews*, and once again Dr. Lutchman and Dr. Akula provide a very significant discussion around this critical PSM element. The chapter provides a very practical approach towards this subject which is a very important element on the PSM system in preventing major incidents. PSSR considerations, the roles and responsibilities of the multidisciplinary commissioning teams, the team composition, the risk management approaches, and the execution of PSSR as well as the generation, review, and approval of the PSSR report and corrective actions is also discussed.

The same two authors deliver through Chapter 13 a focused and comprehensive text on contractor management. Once again, in a very pragmatic way the contractor management element is addressed, explaining the drivers starting from the contract lifecycle. This is an important subject as in the oil and gas, and process industries much of the working hours are undertaken by contractors. Managing their safety has clear benefits on an organization. The chapter addresses human factors and managing the contractors through a contractor management strategy and system. The chapter goes through the contractor prequalification, evaluation, selection, and risk ranking of contractors, and there is a very good discussion of the HSE alignment workshops. There is also a good discussion around leadership visibility.

Chapter 14 is delivered by one of the most renowned practitioners and leaders of fire and safety in the Middle East, Ahmed Khalil Ebrahim. With nearly four decades of hands on experience in emergency management and control, Ahmed provides through this chapter a comprehensive review of emergency and crisis management. The chapter comprehensively describes emergency response planning and management in general terms; provides the basis of the international incident management system and the incident command system (ICS); explains the philosophy of incident classification and how scalable response can be initiated.

The chapter also pragmatically explains how the response organizations are structured as well as the mechanisms utilized to trigger activation of a response; the incident management processes and more significantly in today's organizational resilience approaches, the linkages to business continuity and the enterprise risk management frameworks. The chapter also covered effectively the mutual aid aspect of the emergency and dealing with external stakeholders. ERP is a fundamental element and an integral part of any PSM system for any organization, and Ahmed succeeded to bring this to light in this critical chapter.

In Chapter 15, Jannette Edmonds returns to provide an insightful discussion on human performance within *PSM Compliance Assurance*. In this chapter, a discussion on the balance between technical HAZOP studies and the importance of addressing the human factors is provided with such purpose and clarity and provides an excellent learning to any practitioner, be they a PSM specialist or a human factors specialist. This chapter contributes greatly to the very heart of the subject of this text book. Some of the central themes and concepts which are introduced in Chapter 2 are reiterated and reinforced, but here the value addition comes from the practical application of this knowledge. She also introduces the aspects relating to performance shaping factors (PSFs) or performance influencing factors (PIFs) and then later after explaining the gaps in PSM assurance provides a comprehensive human performance assurance methodology.

Following from this, Chapter 16 addresses the subject *Regulating PSM and the Impact of Effectiveness*. Explaining the purpose of regulations briefly and then a practical discussion on the differences and impacts/limitations/differences between prescriptive and performance-based regulations from a PSM perspective.

In Chapter 17 the authors Dr. Lutchman and Dr. Akula discuss how to prepare an organization for the implementation and alignment of the PSM across the organization. They address the importance of understanding how to transition PSM from an immature to a sustainable state within the company's operating systems and processes. Practitioners would hopefully gain greatly from understanding better the organizational change management requirements and how to create stakeholder alignment, buy-in, and commitment necessary for implementing PSM across the organization. It must be said that this chapter provides a great practical recipe for practitioners who are embarking on creating or otherwise transforming the organizational PSM systems.

Ram Goyal returns in Chapter 18 to provide a great chapter on learning from incidents. With a very succinct and professional review of incidents, Ram brings about the importance of PSM in the understanding of why incidents happen, and more importantly how they can be thus prevented. Through his practitioner's insights, this chapter provides excellent case studies that must be read extremely carefully. What must be said is significant in this chapter is the discussion on the barriers to learning, or what some commentators called the "antilearning mechanisms." He concludes the chapter with very good recommendations of how to learn from incidents effectively.

In Chapter 19, Dr. Mark Hodgkinson provides a review of measuring the effectiveness of implementation. With practical experience from the implementation of operational excellence, Mark explains in details how to gauge the effectiveness of implementation and measure the performance of PSM activities. The chapter can help practitioners in addressing how to prove that the PSM system is in place and delivering the kind of added value and effective loss prevention which was intended for it.

The final Chapter 20 is a reflective discussion from a practitioner and technical specialist who has also worked as a regulator for the oil and gas sector. The chapter addresses human errors, organization culture, and leadership, and links the human elements with PSM and the organizational culture, which he maintains is driven by the leadership. It is a great chapter to conclude this book with, and it was one of the last chapters to have been written and edited in this book.

As the reader will appreciate, PSM and human factors are very much interlinked, and while there has been a great deal of work done to date in many areas of safety, process safety, human performance, and human elements, after reading this book, a practitioner will come to appreciate the importance of understanding the significance of managing systems and managing people.

As a practitioner, myself who has worked closely with systems, conducted both field work studies and semiacademic research in areas including successful implementation of systems, safety cultures, leadership, and learning; this book has provided me personally with such insights and learning, which I could not have otherwise planned for. The kind of practical knowledge and insights that each author has brought through their specialties is fantastic—a diverse mix of topics and approaches to subjects that for years, for myself as a practitioner did not appreciate the strength of the relationship with.

As I was writing and editing this book, I reflected on some of my old action-based research and presentations made in various conferences. In 2008, in a process safety conference I made a presentation on the "Comparative Study between Process Based Safety Management Systems and Integrated (EHS) Safety Management Systems for the Oil and Gas Industry: Strengths and Weaknesses" and in which at the time I believed and argued that besides others, one of the key gaps in PSM systems was competence assurance. It was the human element which I felt had been missing. Then in 2009, I published a paper on the "Enhancing Company Operational Performance through effectively engaging Safety Culture within the Emirates National Oil Companies Distribution Operations, United Arab Emirates," which was case study of a study undertaken to try to understand the challenges to overcome to help the leaders within the organization drive a more effective safety culture. But it was in 2015, when I was invited to an operational excellence gathering of minds of one of the global oil majors where during a discussion on effectiveness of HAZOPs that I came to realize that the human element was one of the most important and the missing parts of our HAZOP practice. Later, in the same year we had arranged for more than 20 different persons from all disciplines and operations to attend a certified program on human factors. The plan was to then use these same persons, now trained in human factors as HAZOP team members.

Reflecting today, two years ago when I decided to produce this book, and realizing that all my experiences, exposure, and research was insufficient to do this subject of PSM and human factors justice, I leveraged my professional network and working relationships with those who became the authors of the chapters of this book, to attempt to produce a practical and authoritative guide to practitioners who were struggling with this subject. This book focuses on providing a practitioner's approach guide—a guide which I would have greatly benefited from as a developing practitioner some 15 years ago and therefore wished had existed at the time.

The this end, I must say in closing that whilst this project has been for me a formidable task that has taken me more than 18 months with my colleagues, it has been a very humbling learning experience, which words cannot describe how much I have enjoyed. I therefore am grateful to every single contributor for their efforts, hard work, passion, and most importantly the faith they placed in me to deliver this purposeful text.

I hope that all practitioners will not only gain from this book and refer to it to guide them in their journey to success, but that they will enjoy reading it and be inspired with the passion it was written with.

Dr. Waddah S. Ghanem Al Hashmi,
Chief Editor

Sample PSSR checklist and report

Appendix 1: Sample PSSR Checklist and Report
PRESTART-UP SAFETY REVIEW

Description of asset reviewed:

_____

_____

1. APPROVAL/AUTHORIZATION

The undersigned confirms (to the best of his / her knowledge) that the assets described above are safe to start-up upon completion of those prerequisite items listed on the following page.

ORIGINATOR (Name, title, date):

Name	
Title	
Date	

Subject Matter Expert (Name, title, date):

	HSE	Mechanical	Technical	Operations	Other
Name					
Title					
Date					

PRODUCTION AREA SUPERVISOR/ UNIT OPERATIONS LEADER (Name, title, date):

_____

Name	
Title	
Date	

PHA REQUIRED? ☐ Yes ☐ No _____ (Responsible individual initials) (If Yes, attach a copy of PHA)

Action Items That Shall Be Completed Prior to Start-up

This equipment has been inspected and found acceptable for start-up after completing <u>all</u> action items listed below. These items present significant risks of major incidents and may significantly impact production if not addressed prior to startup. These are considered A- deficiencies

Action Items Required Prior To Start-up	Implementation Responsibility	Date/Initial when Completed

Corrective Actions Allowed After Start-up With Finite Completion Dates

The following corrective actions have been identified for implementation. Considered B-Deficiencies, they pose no safety nor process safety risk. Nor do they significantly impact production if not completed prior to the start-up of the equipment.

Additional Follow-up Items	Implementation Responsibility	Date and Sign When Completed

Information of Process Unit/Facility Being Added / Modified / Start-up

The following information should be identified and reviewed with the inspection team prior to completing the field assessment to ensure that all team members are aware of known hazards.

Process Materials	Material Hazards

Operating Conditions

Temp.:	
Electrical Classification of Area:	
Special Hazards:	
Environmental Hazards:	
Other	

Documentation Items

Ref*	Category	Apply (Y/N)	O.K.? (Y/N)	Req'd for Startup (Y/N)	Remarks
	Environmental Permit Approved				
	Test Authorization Approved				
	Process safety-critical Equipment, Lines and Interlocks Identified and Documented?				
	Operation Procedure Updated and Issued				
	Process and Instrument Diagram Updated and Accurate				
	Process Hazard Analysis Completed and Results Communicated				
	Process Hazard Analysis Startup Recommendations Complete				
	Technical Standards Updated				
	Operations Reading Sheets Updated				
	Chemical/Vessel Cross-Reference Charts Updated				
	Change in Maximum Intended Inventory of Chemicals Documented				

Ref*	Category	Apply (Y/N)	O.K.? (Y/N)	Req'd for Startup (Y/N)	Remarks
	Update Risk Management Plan [RMP] (changes to storage quantities of risk management substances)				
	Design Basis and/or Design Standards Updated for Process safety-critical Equipment				

Piping and Equipment

Ref*	Category	Apply (Y/N)	O.K.? (Y/N)	Req'd for Startup (Y/N)	Remarks
	Proper Piping and Flanges				
	Correct Gaskets and Bolting. No Short Bolting				
	Installed In Accordance With Pipe Specifications				
	Supports/Vibration Isolation				
	Correct Insulation				
	Tracing Completed				
	Proper Isolation Valves				
	Valve Handles Installed				
	Correctly				

Ref* Category	Apply (Y/N)	O.K.? (Y/N)	Req'd for Startup (Y/N)	Remarks
Low Point Drains/Flush Points OK				
Out-Of-Service Equipment Removed				
Process To Service Connections Comply with Applicable Requirements				
Pressure Test Complete				
Relief Devices Installed and Fully Documented in Critical Components				
Critical Components Inspection Traveler Completed for Initial Inspection and Installation				
Valve Shields Required?				
3-Way Valve Orientation Marked				
Correct Materials of Construction				
Quality Assurance Checks performed, if Quality Assurance Listed Service				
Critical Piping/Valves Identified and in Critical Components & Field Tags in Place.				
Temporary Hangers Removed				
Are Restraining Cables Installed on Chain Operated Valves?				

* Insert local references as appropriate.

Instrumentation

Ref*	Category	Apply (Y/N)	O.K.? (Y/N)	Req'd for Startup (Y/N)	Remarks
	Loop Sheets Updated				
	Pressure Gauges As Required				
	Temperature Gauges As Required				
	Fail-Safe Valve Positions				
	Loop Check Completed				
	Analyzer Check Completed				
	Spare Parts in Stock and/or Spare Parts Listed in CMMS				
	Radiation Sources Check And Approved By Radiation Officer				

Mechanical Equipment

Ref*	Category	Apply (Y/N)	O.K.? (Y/N)	Req'd for Startup (Y/N)	Remarks
	Guards Installed				
	Lubrication/Alignment Complete				
	Correct Rotation				
	Preventive Maintenance Plan In Place				

Ref*	Category	Apply (Y/N)	O.K.? (Y/N)	Req'd for Startup (Y/N)	Remarks
	Spare Parts Set-Up and/or Spare Parts listed in Appropriate Database				
	Hydraulic Systems Tested				
	Area Equipment Files Updated				
	Proper Access To Equipment				

* Insert local references as appropriate.

Electrical Equipment

Ref*	Category	Apply (Y/N)	O.K.? (Y/N)	Req'd for Startup (Y/N)	Remarks
	Conforms to Elect. Classification				
	Grounding and Bonding Complete and Tested				
	Lightning/Surge Protection Is Complete				
	No Open Conduits				
	Out-Of-Service Equipment Removed				
	Preventive Maintenance Plan In Place				
	Process/Safety Interlocks Tested and Documented				
	Update Lighting and Power Drawings				

Structures

Ref*	Category	Apply (Y/N)	O.K.? (Y/N)	Req'd for Startup (Y/N)	Remarks
	Fire Doors and Fire Walls				
	Doors and Ramps				
	Traffic Flow OK				
	Emergency Exits Clear/Adequate				
	Handrails, Guardrails and Safety Chains Adequate and In Place				
	Ladders and Stairs				
	Walking Surfaces Adequate				
	Grating Anchored				
	Floor Drainage, Curbs, Gutters and Dikes				
	Overhead Hazards Removed				
	Operating and Maintenance Access Adequate				
	Lighting OK Day and Night				
	Safety Showers And Eyewash Stations Adequate				

* Insert local references as appropriate.

Procedures

Ref*	Category	Apply (Y/N)	O.K.? (Y/N)	Req'd for Startup (Y/N)	Remarks
	Ventilation System Checked				
	PA System/Radios Adequate				
	Emergency Procedures Adequate				
	Spill Procedures Adequate				
	Material Data Safety Sheets (MSDS) Approved				
	Instrument Calibration Procedures				
	Analyzer Calibration Procedures				
	Maintenance Procedures in Place				
	Vessel Entry Set-up List Complete				
	Rescue Plans for Vessels Updated				
	Personal Protective Equipment (PPE) Chart Updated				
	Operator Training Plans Updated				
	Operator Training Complete and Documented				
	Mechanic Training Complete and Documented				
	Training for Supervisors and Professionals Complete and Documented				

* Insert local references as appropriate.

Labelling

Ref*	Category	Apply (Y/N)	O.K.? (Y/N)	Req'd for Startup (Y/N)	Remarks
	Lines Labeled for Content and Flow Direction				
	Instrument Labeled				
	Switches and Pushbuttons				
	High Voltage Marked / Guarded				
	Hot Surfaces Marked / Guarded				
	Insulation Identified				
	All Equipment Labeled				
	Have Motor Control Centers (MCC) been Labeled On Incoming Power Section				
	Emergency Exits Marked				
	Monorails Marked With Load Limit				
	Utilities Stations and Services Labeled				
	Hearing Protection Areas Identified and Clearly Marked				
	PPE Areas Identified and Clearly Marked				

Samples

Ref*	Category	Apply (Y/N)	O.K.? (Y/N)	Req'd for Startup (Y/N)	Remarks
	Sample Method Developed				
	Plans For Disposal of Sample In Place				
	Sample Schedule Updated				

*Insert local references as appropriate.

Ergonomics

Ref*	Category	Apply (Y/N)	O.K.? (Y/N)	Req'd for Startup (Y/N)	Remarks
	Has the Facility Been Constructed So That the Need for Stooping, Bending, Stretching, Over Reaching and Work Overhead During Operation Has Been Eliminated or Reduced to Minimum?				

	Ergonomic Considerations – *See PHA Change of Design Checklist*				
	Has Need to Lift, Carry, Push or Pull Heavy Loads or Parts Been Minimized?				
	Visual Display Screens Free From Glare				

General Items

Ref*	Category	Apply (Y/N)	O.K.? (Y/N)	Req'd for Startup (Y/N)	Remarks
	Tripping Hazards and Pinch Points				
	Fire Protection and Sprinkler Systems				
	Fire Extinguishers				
	Self-Contained Breathing Apparatus				
	Emergency Lighting				
	Hoses Inspected (both Chemical & Utility)				
	Noise Monitoring Needed				
	Respirators Available if Required				
	Special Tools Available if Required				
	Are Hoists, Trolleys and Monorails Inspected and Added to Inspection File?				
	Construction is in accordance with design specifications.				
	Hidden/Stored Energy Sources Identified				
	Required Personal Protective Equipment (PPE) Available				
	Safety Training Complete				

	Technical Competency Training Completed				
	Painting Complete				
	Area Meets Housekeeping Standards				
	Do Any of These Changes Require Contractor Safety Training Package Updates?				

*Insert local references as appropriate.

2

Reference list and international standards and codes

A comprehensive list of standards, codes, guidance, and regulations relevant to human factors and ergonomics in process safety is given below.

Barrier Management and Safety Critical Tasks

- Human Factors in Barrier Management—CIEHF White Paper
- Bow Ties in Risk Management. A Concept Book for Process Safety. CCPS in association with the Energy Institute
- HSE OTO 1999/092—Offshore Technology Report—Human Factors Assessment of Safety Critical Tasks
- Guidance on Human Factors Safety Critical Task Analysis. Energy Institute
- Process Safety—Recommended Practice on Key Performance Indicators. IOGP 456

Cognitive/behavioral performance managing human performance and failures, effective communications, and human factors in design

- HSG 48—Reducing error and influencing behavior
- HSG 263—Involving the Workforce in H&S
- HSG 218—Stress Management
- ISO 9241–11:2018 Ergonomics of human-system interaction—Part 11: Usability: Definitions and concepts
- ISO 9241–125:2017 Ergonomics of human-system interaction—Part 125: Guidance on visual presentation of information
- ISO 11428:1996 Ergonomics—Visual danger signals—General requirements, design and testing
- ISO 11429:1996 Ergonomics—System of auditory and visual danger and information signals
- L64—Safety Signs and Signals

- ISO/CD TS 9241-126 [Under development] Ergonomics of human-system interaction—Part 126: Guidance on the presentation of auditory information
- ISO 9921:2003 Ergonomics—Assessment of speech communication
- ISO 10075–1:2017 Ergonomic principles related to mental workload—Part 1: General issues and concepts, terms and definitions (Parts 1–3)
- ISO/TR 22100–3:2016 Safety of machinery—Relationship with ISO 12100—Part 3: Implementation of ergonomic principles in safety standards
- ISO 27500:2016 The human-centred organization—Rationale and general principles
- ISO 27501:2019 The human-centred organization—Guidance for managers
- Designing for Human Reliability. R.M. McLeod. Elsevier

Systems involving human intervention and controls managing human, effective communications, human factors in design

- L111—The Control of Major Accident Hazards Regulations 2015 (COMAH)
- COMAH CA—Inspection of Electrical, Control and Instrumentation Systems at COMAH Establishments
- ISO 6385:2016 Ergonomics principles in the design of work systems
- BS EN ISO 11064-1 2001 Ergonomic design of control centres—Part 1 Principles for the design of control centres (Parts 1–7)
- EEMUA 191—Alarms Management
- BS EN ISO 9241-210 2010 Ergonomics of human-system interaction. Human-centred design for interactive systems
- ISO 9241–2:1992 Ergonomic requirements for office work with visual display terminals (VDTs)—Part 2: Guidance on task requirements
- ISO 9241–500:2018 Ergonomics of human-system interaction—Part 500: Ergonomic principles for the design and evaluation of environments of interactive systems
- ISO/TR 16982:2002 Ergonomics of human-system interaction—Usability methods supporting human-centred design
- ISO/TS 18152:2010 Ergonomics of human-system interaction—Specification for the process assessment of human-system issues
- CRR 74/1998—Developing a framework for Participatory Ergonomics (HSE)

Physical spaces: human performances and failures, human factors in design

- BS EN 527-1 2011 Office Furniture—Work tables and desks—Part 1 Dimensions

- HSG57—SEATING
- ISO 9241–5:1998 Ergonomic requirements for office work with visual display terminals (VDTs)—Part 5: Workstation layout and postural requirements
- BS EN 14056-2003 Laboratory furniture. Recommendations for design and installation
- BS EN 894-1 A1 2008 Safety of machinery. Ergonomics requirements for the design of displays and control actuators. General principles for human interactions with (Parts 1–4)
- ISO 9355–1:1999 Ergonomic requirements for the design of displays and control actuators—Part 1: Human interactions with displays and control actuators (Parts 1–4)
- ISO 9241–400:2007 Ergonomics of human-system interaction—Part 400: Principles and requirements for physical input devices
- BS EN ISO 14122-1 2001 Safety of machinery. Permanent means of access to machinery. Choice of a fixed means of access between two levels (Parts 1–4) (STEPS, STAIRS, LADDERS)
- BS EN 131-1 2007 + A1 2011 Ladders. Terms, types, functional sizes (Parts 1–4)
- BS EN 12464-1 2002 Light and lighting. Lighting of work places. Indoor work places
- ISO/CIE 8995–3:2018 Lighting of work places—Part 3: Lighting requirements for safety and security of outdoor work places
- BS EN ISO 24502 2010 Ergonomics. Accessible design. Specification of age-related luminance contrast for coloured light
- ISO 14738:2002 Safety of machinery—Anthropometric requirements for the design of workstations at machinery
- ISO 15534-1:2000 Ergonomic design for the safety of machinery—Part 1: Principles for determining the dimensions required for openings for whole-body access into machinery (Parts 1–3)

Physical/health environment human performances and failures, human factors in design

- ISO 26800:2011 Ergonomics—General approach, principles and concepts
- ISO 45001—Occupational health and safety
- OHSAS 18001:2007 Occupational health and safety management systems—Requirements
- L153—Managing health and safety in construction—Construction (Design and Management) Regulations 2015 (CDM)
- L23—MH Regs AND ISO 11228–1:2003 Ergonomics—Manual handling—Part 1: Lifting and carrying (Parts 1–3 for lowering, pushing, pulling and high freq loads)
- ISO 11226:2000 Ergonomics—Evaluation of static working postures

- ISO/TS 20646:2014 Ergonomics guidelines for the optimization of musculoskeletal workload
- L25—PPE Regs
- L26—DSE Regs
- L108—NOISE AND VIBRATION
- HSG 60—UPPER LIMBS
- INDG 401—Working at Height 2005
- ISO 7250–1:2017 Basic human body measurements for technological design—Part 1: Body measurement definitions and landmarks
- ISO 9241–6:1999 Ergonomic requirements for office work with visual display terminals (VDTs) —Part 6: Guidance on the work environment
- ISO 9241–392:2015 Ergonomics of human-system interaction—Part 392: Ergonomic recommendations for the reduction of visual fatigue from stereoscopic images
- Thermal Environment—Many, many standards (measurement, definitions, effects of clothing, equipment, activity, exposure etc.) INCL
- ISO 11399:1995 Ergonomics of the thermal environment—Principles and application of relevant International Standards

Index

Note: Page numbers followed by "*f*," "*t*," and "*b*" refer to figures, tables and boxes, respectively.

Printed in the United States
By Bookmasters